SUSTAINED

SUSTAINED

CREATING A SUSTAINABLE HOUSE THROUGH SMALL CHANGES, MONEY-SAVING HABITS, AND NATURAL SOLUTIONS

CANDICE BATISTA

Coral Gables, FL

Copyright © 2024 by Candice Batista.
Published by Mango Publishing, a division of Mango Publishing Group, Inc.

Cover & Layout Design: Megan Werner
Cover Art: elinacious, tanawatpontchour / stock.adobe.com
Interior Art: Abdie, Ekaterina, Liubov, nadiinko, Onur Cem, Rawpixel.com, Tasha Vector / stock.adobe.com

Mango is an active supporter of authors' rights to free speech and artistic expression in their books. The purpose of copyright is to encourage authors to produce exceptional works that enrich our culture and our open society.

Uploading or distributing photos, scans or any content from this book without prior permission is theft of the author's intellectual property. Please honor the author's work as you would your own. Thank you in advance for respecting our author's rights.

For permission requests, please contact the publisher at:
Mango Publishing Group
2850 S Douglas Road, 2nd Floor
Coral Gables, FL 33134 USA
info@mango.bz

For special orders, quantity sales, course adoptions and corporate sales, please email the publisher at sales@mango.bz. For trade and wholesale sales, please contact Ingram Publisher Services at customer.service@ingramcontent.com or +1.800.509.4887.

Sustained: Creating a Sustainable House Through Small Changes, Money-Saving Habits, and Natural Solutions

Library of Congress Cataloging-in-Publication number: 2023947328
ISBN: (hc) 978-1-68481-350-6 (pb) 978-1-68481-351-3 (e) 978-1-68481-352-0
BISAC category code: HOM022000, HOUSE & HOME / Sustainable Living

The information provided in this book is based on the research, insights, and experiences of the author. Every effort has been made to provide accurate and up-to-date information; however, neither the author nor the publisher warrants the information provided is free of factual error. This book is not intended to diagnose, treat, or cure any medical condition or disease, nor is it intended as a substitute for professional medical care. All matters regarding your health should be supervised by a qualified healthcare professional. The author and publisher disclaim all liability for any adverse effects arising out of or relating to the use or application of the information or advice provided in this book.

Dedicated to Planet Earth and its entire cast of characters—from people to penguins—because, let's face it, we're all in this together!

CONTENTS

Introduction	8
Chapter 1—From Independence to Interdependence	13
Chapter 2—The Conscious Kitchen: Food Waste, Climate Change, and You	25
Chapter 3—Squeaky Green: Navigating the World of Natural Cleaners	61
Chapter 4—From Suds to Sustainability: Rethinking Your Laundry Routine	93
Chapter 5—Sewing Seeds of Change: Your Guide to Ethical Fashion	119
Chapter 6—Skin Deep: The Real Impact of Sustainable Beauty	155
Chapter 7—Bathroom Revolution: From Waste-Land to Waste-Free	185
Chapter 8—Nesting Naturally: Rejecting Fast Furniture for a Sustainable Home	211
Conclusion—Tipping Point: The Urgent Need for Sustainable Choices	239
Acknowledgments	243
About the Author	244
Bibliography	245

Introduction

In the ecological and environmental context, "sustained" refers to the enduring and balanced approach to economic activity, social progress, and environmental responsibility. It embodies the capacity of ecosystems to endure, emphasizing the interconnectedness of our natural systems and our responsibility to future generations.

When we speak of a "sustained" environment, we mean a healthy and vibrant long-term habitat that maintains its biodiversity, productivity, and the processes that drive it. This involves conserving resources, supporting a diversity of species, and preserving the natural processes that keep ecosystems functioning. It's about utilizing resources in such a way that they're not depleted or permanently damaged.

The concept of being "sustained" in an environmental context also extends to human societies and their economic systems. It advocates for a balanced approach considering economic growth, social equity, and environmental protection. Therefore, sustained economies operate in harmony with the environment, promoting the wise use of resources and reducing waste and pollution.

"Sustained" is inherently a forward-looking concept. It represents a commitment to leaving a planet that future generations can live in and enjoy, just as we have. It's an acknowledgment of our responsibility to the earth and each other, as we all share the benefits of a sustained and healthy planet.

Growing up in South Africa, I was fortunate to have a deep connection to the natural world from an early age. There was nothing quite like the excitement of embarking on a safari and encountering magnificent animals in their natural environment. Nature's breathtaking beauty and power left an indelible impression on me, fostering a sense of wonder and gratitude that has remained with me. As I lay beneath the starlit sky and listened to the sounds of the wilderness, I felt an intense connection to our planet. I understood that we are not separate from nature but an integral part of it. As I got older, my connection to the natural world grew stronger, leading me to recognize the significance of living sustainably and nurturing the planet that supports us.

However, as I became more knowledgeable about the environment, I became increasingly aware of the darker aspects of our relationship with nature. I was particularly struck by the prevalence of environmental racism, a form of institutional racism that disproportionately impacts communities of color by situating landfills, incinerators, and hazardous waste disposal sites in their neighborhoods. Witnessing the catastrophic effects of this discrimination ignited a passion within me.

Environmental justice is a dynamic concept that has assumed unique meanings in different parts of the world, and South Africa is no exception. The idea, which originated in the United States fifty years ago in response to the discriminatory practices identified by Black communities as "environmental racism," made its way to South Africa during a conference of environmental activists in 1992. At the time, combating environmental racism was considered essential to democratization, particularly between 1992 and 1994. Today, however, a form of environmental racism persists, as most Black South Africans continue to reside on the most degraded land, in the most polluted neighborhoods near coal-fired power plants, steel mills, incinerators, and waste sites. Many lack access to clean air, water, and essential services. Climate change exacerbates these issues, with devastating consequences for people experiencing poverty and the working class, as evidenced by soaring food prices, crop failures, and water scarcity.

As an environmental journalist, I have dedicated my career to covering stories about global warming, ecological stewardship, and the fight for environmental justice. Over my thirty-year career, I have had the privilege of meeting and interviewing individuals worldwide working to protect the environment and combat environmental racism. I have visited communities affected by climate change, from rising sea levels to devastating wildfires, witnessing firsthand the impact of our changing climate on vulnerable populations.

To raise awareness about these pressing issues, I have reported on innovative technologies that can help reduce our carbon footprint, the significance of sustainable agriculture and conservation, and the inspiring endeavors of local activists and national governments to address environmental injustice. Yet, despite the many challenges we face, I remain hopeful. I have seen the power of individuals and communities to enact positive change.

Sustained reflects my passion for journalism and commitment to the environment. I hope my journey inspires you in some way. Remember, Rome was not built in a day. Change takes time! It's a process. Ralph Waldo Emerson's famous quote, "Life is a journey, not a destination," encapsulates the mindset necessary for pursuing a *sustained* lifestyle in a world that often promotes the opposite.

It's not about being perfect. But it is about taking action and owning up to the realities of the waste we are creating and what we can do about it.

I will share my approach to sustainability and mindful living throughout this book. At its core, it's about understanding your impact on the world. It's the connection I refer to above and in Chapter 1.

On my website, The Eco Hub, I created a set of criteria for suggesting brands or products. It's not perfect, but it gives you a guideline. You can use these as you see fit. Maybe one is more important than the other to you.

Over the years, these have aided me in navigating the challenging world of living with less waste! Or trying to, at the very least.

MY ETHICAL CRITERIA

Criteria 1: What Is It Made From?

When we talk about products, especially in a world where everything feels so instant, it's crucial to take a *sustained* pause—ever held a soft cotton shirt against your skin? Feels dreamy, right? But did you know that behind that simple pleasure lies a complex tapestry of decisions that impact our planet?

Consider cotton. It's nature's gift, but how we grow it can tax our resources. But then there's organic cotton—the superhero version. It uses less water and ensures Mother Earth isn't doused with chemicals.

Beyond fashion, think about your favorite wooden coffee table or that elegant gold bracelet that catches the light just right. These items have stories too. Are they tales of sustainable harvesting and ethical mining, or are they narratives of exploitation? We can ensure a harmonious relationship with our environment by choosing items with a story that aligns with the principles of being *sustained*.

And, oh, that skincare regime you swear by? Look a tad closer. We want those ingredients to nourish our skin, but they should also sustain our planet. No one looks good wearing unsustainably sourced palm oil.

Criteria 2: Sustainable Sourcing/ Ethical Manufacturing

We live in an age of hashtags. #WhoMadeMyClothes became a rallying cry, but why stop there? Extend it—#WhoMadeMyShoes, #WhoMadeMySkincare, even #WhoMadeMyCoffeeBeans.

It's been a decade since the Rana Plaza tragedy. A sobering reminder that every tag, stitch, and product embodies real hands and lives. When we choose to live *sustained*, we don't just ask these questions; we demand answers. We insist on knowing the human stories behind our products.

Have you ever tried darning a sock? I did once; I felt like I was defusing a bomb with all the precision it needed! Someone, somewhere, is expertly weaving, stitching, and crafting for us. It's only fair we ensure their world is as *sustained* as we want ours to be.

Criteria 3: Corporate Social Responsibility

In today's rapidly changing climate, a brand's commitment to carbon reduction is commendable and imperative. Regardless of its sustainability badge, every brand should transparently declare its strategies for carbon trimming, showcasing tangible yearly reductions rather than grand but hollow promises. Programs like carbon offsets, though commendable, need rigorous evidence to back them up. Brands like tentree have set respectable benchmarks, but we need more players to champion Patagonia-level ambitions. Beyond carbon, sustainability transcends to other facets of business: creating quality, long-lasting products; envisioning a product's life cycle beyond the consumer, which could mean repair, resale, or recycling; championing handcrafted items not only to reduce machinery emissions but also preserving age-old crafting techniques; adeptly managing resources from water to energy; ensuring even the corporate office breathes the ethos of sustainability through waste minimization, recycling, and composting; evolving packaging that's both eco-friendly and reusable; and making transport decisions that prioritize the environment. When a company gets certified, like through LEED, it's a testament to its all-encompassing pledge to our planet. Truly sustainable corporate responsibility doesn't stop at products; it's a holistic approach that infuses the *sustained* philosophy into every facet of the business.

Phew! That's *a lot* for the average person to take in, I know! But it's the reality of where we are right now. So let's try and fix it. The suggestions above are based on my experiences and the success stories of hundreds of people I've encountered on this journey.

By the end of this book, you'll be empowered with the knowledge to create a more sustainable and eco-friendly life while being mindful of the stuff you use daily.

Plus, with my cost breakdowns comparing unsustainable items to sustainable alternatives, you'll be amazed by the significant savings you can achieve by going green. Keep a look out for this symbol.

When you see it, you know you are saving money.

Let's embark on this journey together and learn to embrace the joys of minimalism, making Mother Earth proud as we develop our newfound ability to resist the allure of unnecessary possessions. After all, the best things in life aren't things! By consciously opting for a *sustained* lifestyle, you're not just making ripples but waves of positive change. So, next time you shop, eat, or even gift, remember the *sustained* ethos. It's more than just a word; it's a way of life.

CHAPTER 1
From Independence to Interdependence

As I embark on the journey of writing this book, I reflect on the world we live in, the choices we make, and the impact of those choices on our planet. Of course, I've always been conscious of the need to live sustainably. Still, when I started delving deeper into the concepts of circular and linear economies, I realized how much our current system is at odds with the ideals of sustainability and frugality.

At its core, a linear economy follows the extract-produce-distribute-consume-dispose pattern. It is a one-way street that relies heavily on extracting finite resources from the earth, producing goods using those resources, distributing them to consumers, and ultimately discarding them as waste when they are no longer helpful. In this linear model, our economy is built on the assumption of endless growth and consumption, with little regard for the environmental and social consequences.

In contrast, a circular economy seeks to break this pattern by designing products and systems that minimize waste, reduce the need for resource extraction, and ultimately regenerate natural systems. The circular economy strives to create a world where waste is minimized, and resources are used in a closed-loop system by promoting product longevity, renewability, reuse, and repair.

As I delved into these concepts, I thought about the role of planned obsolescence in our linear economy. Planned obsolescence is designing products with a limited lifespan, often intentionally, to encourage consumers to replace them more frequently.

Unfortunately, this strategy drives sales and profits for manufacturers and contributes to the mountains of waste in our landfills and oceans.

Everything I was learning made me question my consumer choices and the products I brought into my home. Could I find a way to live sustainably, save money, and reduce my contribution to this wasteful linear system? I knew I couldn't change the world overnight, but I believed every small step could make a difference.

HOW DOES "ZERO WASTE" FIT INTO ALL OF THIS?

The zero-waste movement began as a response to the unsustainable practices of the linear economy, aiming to minimize waste generation and promote a more circular approach to resource management. Bea Johnson, often called the Queen of Zero Waste, has been a prominent figure in the movement, showcasing her minimal-waste lifestyle with her famous "trash in a jar" approach. While Johnson's methods may be inspiring, they may seem unattainable for many people, making zero waste overwhelming.

It's essential to recognize that sustainable living is challenging in a consumer-driven society, where waste generation is the norm. With many trash cans in our homes and surroundings, it's no surprise that the average American generates four and a half pounds of landfill-bound garbage daily. To better understand the concept of zero waste, we need to examine how landfills operate and the problems they pose.

Contrary to popular belief, landfills are not giant compost piles. Instead, they contain various types of waste, from food and hazardous materials to plastics and paper. As more waste is added, the items below become deprived of oxygen and sunlight, hindering decomposition. Studies have found things in perfect condition buried in landfills dating back to the 1940s.

We must address two significant concerns with landfills: leachate and methane gas. Leachate is a significant issue for municipal solid waste (MSW) landfills, posing a severe threat to surface water and groundwater.

It is defined as the toxic liquid that forms when precipitation enters the landfill, mixing with the moisture present in the waste. This poisonous mixture accumulates at the bottom of the landfill, leaching into the soil and contaminating groundwater.

Methane gas is another problem associated with landfills. As a potent greenhouse gas, methane traps heat in our atmosphere, contributing to global warming. Dumps are one of the largest sources of methane emissions, primarily due to the decomposition of organic materials, such as food waste, in anaerobic conditions (i.e., without oxygen).

The leachate and methane gas issues highlight the urgency of adopting a more *sustained* approach to waste management, such as zero-waste principles. By reducing the amount of waste sent to landfills and promoting the responsible use of resources, we can mitigate these environmental threats and work toward a cleaner, healthier planet.

Another major issue associated with landfills is the presence of plastic. Unlike organic materials, plastic does not biodegrade; it photodegrades, breaking into progressively smaller pieces over hundreds of years. This process contributes to soil and water pollution and the broader issue of plastic waste in our environment.

While useful in theory, landfills have become a significant concern due to our society's overconsumption and reliance on single-use items. In comparison, in the 1940s, people lived relatively trash-free lives before the widespread adoption of single-use products. The growing problems associated with landfills, including leachate, methane gas emissions, and plastic pollution, emphasize the importance of embracing zero-waste principles.

Today, achieving absolute zero waste is nearly impossible. Even sustainable products produce some waste at their source, so "low waste" is often used as a more realistic alternative. To truly transition to a circular economy, we must overhaul our entire system, moving away from industrial production and its reliance on fossil fuels and the take-make-dispose model. Instead, we should adopt innovative techniques prioritizing product longevity, renewability, reuse, and repair. By focusing on these, we can develop products designed for recovery (e.g., beeswax wrap) rather than disposal (e.g., Saran Wrap) that are more durable and long-lasting.

HOW CAN WE LIVE A MORE LOW-WASTE LIFE?

It comes down to being connected to the things we buy, understanding their impact on the world, and being mindful of that (remember my ethical criteria?). Like slow food, defined as "an alternative to fast food, it strives to preserve traditional and regional cuisine and encourages farming of plants, seeds, and livestock characteristic of the local ecosystem." We must apply the same principles of the slow food movement to everything we buy daily.

We might have little choice, especially when it comes to plastic. In 2019, total US plastic waste generation reached 73 million metric tons, equivalent to 221 kilograms of plastic waste per inhabitant. This amount was approximately five times higher than the global per capita average. Despite growing awareness of the environmental impacts of plastic waste, the demand for plastics shows little signs of slowing.

US plastic waste generation is projected to grow to 142 million metric tons by 2060. The staggering growth of plastic waste underscores the urgency of adopting sustainable practices and transitioning toward a circular economy. But, and there is a but, with all things eco, it's never as simple as it seems. Take the debate about what's more eco-friendly, glass or plastic. There's no clear winner—it's a nuanced decision that involves weighing the benefits and drawbacks of each material.

Glass and plastic both have significant environmental impacts. Still, they differ in production, recycling, and health considerations. Glass is made from silica (sand), soda ash, limestone, and dolomite, all-natural raw materials. However, only specific types of sand can be transformed into glass, and it often comes from seabeds and river beds, disrupting natural ecosystems. Glass demands higher heat than plastic, requiring more natural resources, including nonrenewable materials like sand

and minerals, and it releases dangerous particles into the atmosphere, contributing to acidification and smog formation.

On the other hand, glass is 100 percent recyclable and infinitely recyclable, reducing emissions, saving natural materials, and saving energy. For every tonne of recycled glass, roughly 1.2 tonnes of raw materials are saved, cutting water and air pollution by 50 percent and 20 percent, respectively. However, glass recycling still requires a lot of energy, and some virgin materials are still necessary due to losses throughout the recycling supply chain. Additionally, poor recycling rates make the environmental footprint of glass heavy. Only 31.3 percent of the 12.3 million tons of glass generated in the US are recycled.

Conversely, plastic is made from petroleum and natural gas, with monomers joining as building blocks to form various polymers. Plastic requires roughly 8 percent of the world's total oil production. It produces a lot of toxic air pollution, including VOCs, benzenes, toluene, and nitrous oxides, making it one of the most energy-intensive materials. However, plastic is lightweight, making it cheaper to ship, and has a lower carbon footprint for transportation emissions than glass.

Recycling plastic has significant energy savings, reducing barrels of oil, energy, and electricity, but just 9 percent of all the plastic that's ever been produced has been recycled. Even when plastic is recycled, it is transformed into a lesser quality product, and most plastics can only be recycled once or twice. The production and disposal of plastic release harmful microplastics and toxins that can affect human health, including metabolic disorders, fertility problems, and even cancer.

In summary, neither glass nor plastic is a perfect solution. To make a better planetary decision, we should reduce our consumption of both materials, reuse them ourselves, and choose other alternatives such as aluminum, which is light, infinitely recyclable, and requires less energy and resources to produce than both glass and plastic.

Addressing plastic pollution is critical, but combating climate change is an even more urgent priority. Therefore, focusing on more significant concerns, such as reducing meat intake, minimizing food waste, and buying fewer items, is crucial.

We're all familiar with the classic five Rs of zero waste: Refuse, Reduce, Reuse, Recycle, and Rot. However, we can do even more to truly embrace sustainability in our daily lives. That's why I'm proposing an updated set of guiding principles, the nine Rs: Rethink, Refuse, Reduce, Reuse, Repair, Repurpose, Refill, Rot, and Recycle. These principles serve as a valuable roadmap for anyone looking to minimize their environmental footprint and lead a more sustainable lifestyle.

My approach to low waste expands on the traditional five Rs by emphasizing the importance of rethinking our consumption habits, repurposing items, repairing broken items, and refilling containers whenever possible. By prioritizing these additional Rs, we take a more active role in waste prevention and embody the spirit of a circular economy.

By placing "Rot" ahead of "Recycle," we highlight the value of composting and organic waste management. This helps to return essential nutrients to the earth and reduce methane emissions from landfills. However, it's critical to recognize that while recycling is crucial, it shouldn't be our first defense against waste. Instead, our focus should be on the proactive steps we can take to prevent waste and close the loop on resource usage.

Adhering to the nine Rs reminds us to explore every possible waste reduction avenue before recycling. This comprehensive approach aligns with the foundational principles of a circular economy and sets the stage for a genuinely sustainable future.

THE 9 Rs OF ZERO-WASTE

- rethink
- refuse
- reduce
- reuse
- repair
- repurpose
- refill
- rot
- recycle

1. Rethink

The concept of "rethink" in sustainable living and waste reduction encourages you to reevaluate your consumption habits, lifestyle choices, and overall mindset toward resources and waste. It involves questioning the status quo, challenging existing patterns, and considering the environmental impacts of our actions before making decisions.

Rethinking for sustainable living includes:

1. Assessing the necessity of purchases to avoid impulse buys and reduce waste.
2. Choosing durable, repairable, and eco-friendly products for a longer lifespan and reduced environmental impact.
3. Acknowledging the consequences of daily choices, like transportation and energy usage, and opting for more sustainable alternatives.
4. Participating in community-based initiatives, like tool libraries or clothing swaps, to embrace the sharing economy and reduce waste.

2. Refuse

Just say *no*! Learn to be comfortable with refusing things you don't need, like single-use plastics, unnecessary packaging, and items that contribute to waste. By refusing these items in the first place, you set yourself up for success.

Some examples:

1. Avoid taking flyers, pamphlets, or other promotional materials handed out at events or on the street. Most of this information can be found online.
2. Avoid buying pre-cut fruits and vegetables in plastic packaging. Instead, purchase whole produce and cut it at home.
3. Decline disposable amenity kits on flights and bring your reusable travel-size items instead.
4. Use digital loyalty cards or apps on your phone instead of carrying physical cards that can quickly become waste.

3. Reduce

Embrace the concept of letting go. If an item no longer "sparks joy," consider donating it, participating in a clothing swap, or using online marketplaces. Opt for shopping secondhand and focus on purchasing only what you genuinely need rather than what you merely want. Reducing at the source involves thoughtfully considering an item's lifecycle and disposal after use. Remember, one of the main contributors to our waste problem is our addiction to possessions and overconsumption. So, minimize your purchases, like exploring renting options. Make a conscious effort to declutter your home and donate items that are no longer useful.

Avoid feeling overwhelmed when decluttering by focusing on manageable areas like your handbag, a single drawer, or the medicine cabinet. Gradually progress to other spaces in your home, such as the Tupperware drawer. Decluttering is a slow, sustainable process that shouldn't be rushed. By adopting a gradual and mindful approach, you can create lasting changes in your consumption habits and living space.

4. Reuse

Reuse refers to using an item multiple times for its original purpose without significant alterations. Reusing items helps to conserve energy and raw materials that would otherwise be needed to produce new products. Using cloth shopping bags prevents the need for single-use bags, reducing plastic waste. Refilling reusable water bottles helps decrease plastic pollution and energy consumption. Donating or sharing unwanted items prolongs their usefulness. Reusing glass jars for storage reduces the need for disposable containers, and reusing gift bags or wrapping materials prevents waste from new materials.

5. Repair

Repairing items instead of replacing them promotes a more eco-friendly approach and aligns with the principles of a circular economy. When we fix things, we extend their lifespan, reducing the need for new resources to produce replacements. This minimizes waste generation and lessens the demand for energy and raw materials needed for manufacturing.

In a circular economy, the focus is on maximizing the value of products and materials for as long as possible, keeping resources in use, and minimizing waste. Repairing items supports this concept by maintaining the functionality of products, encouraging resource efficiency, and fostering a culture of mindful consumption. By choosing to repair, we also reduce the burden on landfills and decrease pollution generated during the disposal and production of new items.

6. Repurpose

Repurpose refers to creatively adapting, modifying, or transforming an object, material, or resource for a new or different function. This practice extends the life and usability of the item, reducing waste and consumption while promoting sustainability and resourcefulness. Repurposing involves innovative thinking and problem-solving to find new applications for things that might otherwise be discarded or deemed obsolete. Transform old t-shirts, pillowcases, or other fabric items into reusable bags for shopping and reduce the consumption of single-use plastic bags. Instead of throwing away worn-out furniture, refurbish and repurpose it. Paint or reupholster chairs, tables, or dressers to give them a fresh look and extend their life. Use discarded containers, such as tin cans, plastic bottles, or old boots, as planters for small plants or herbs. Repurpose food scraps and other biodegradable waste as compost for your garden, enriching the soil and reducing the need for chemical fertilizers. Salvage wooden pallets and use them to create unique furniture pieces, such as tables, benches, or shelves. Make reusable food wraps from beeswax and fabric scraps to replace single-use plastic wrap or aluminum foil. Mend, alter, or combine old clothing items to create new, unique garments. Reuse empty spray bottles for eco-friendly cleaning from natural ingredients like vinegar, baking soda, and essential oils. Repurpose old building materials, such as bricks, wood, or metal, for new construction projects or DIY home improvements.

7. Refill

Refilling offers numerous advantages, such as reducing waste by reusing containers like water bottles, jars, or cleaning product bottles. This practice significantly reduces single-use packaging waste, helping conserve natural resources and minimize pollution. By promoting a circular economy, refilling keeps resources in use for extended periods and minimizes waste. This approach is also cost-effective, as refilling stations or bulk purchasing options often provide lower per-unit costs, saving consumers money. Lastly, refilling reduces transportation and manufacturing emissions by lowering the energy and resources needed to produce new packaging, decreasing the overall carbon footprint associated with the product.

8. Rot

Composting breaks down organic materials, such as food scraps, yard waste, and other biodegradable items, into nutrient-rich soil amendments.

Microorganisms, worms, and fungi decompose the organic matter through a natural process, ultimately producing a substance called humus. This humus can enrich garden soil, support plant growth, and reduce the need for chemical fertilizers.

By composting organic waste, the concept of "rot" helps reduce the amount of waste sent to landfills, decrease methane emissions from decomposing organic matter, and conserve resources by recycling nutrients back into the soil. We will dive into this topic more in Chapter 2.

9. Recycle

Recycling is a crucial process but should be considered a last resort. Regrettably, only a tiny percentage of waste is recycled, and most recycled materials are down-cycled into lower-quality, disposable items. Instead, prioritize the first eight steps and practice "precycling" by assessing an item's value, usefulness, and recyclability before buying it. Follow local recycling regulations to ensure that recyclable items are handled correctly and not sent to landfills. Compliance with these rules reduces contamination, which occurs when non-recyclable materials are mixed with recyclable ones or when things need to be cleaned before recycling. Supporting local initiatives contributes to better resource management and waste reduction. Proper recycling conserves resources, minimizes pollution, and decreases greenhouse gas emissions. By staying informed, you can educate others and promote responsible recycling habits.

———

You can make significant strides toward living a low-waste lifestyle by following the nine Rs. The goal of adopting these practices is to deepen our understanding of our connection to the planet and one another.

Embracing interdependence rather than independence encourages us to work together and in harmony with nature to create a more sustainable and balanced world for all.

In choosing to focus on the home as the hub for sustainable living, I wanted to emphasize everyone's significant role in fostering a more environmentally responsible world.

Our homes serve as microcosms of the larger world, and the choices we make within them can have vast implications. Supporting fast fashion, for instance, may seem like a matter of personal style or budget, but it contributes to an unsustainable global production and waste system. By being mindful of our choices and their impact on people worldwide, we can adopt eco-friendly habits and educate others about sustainable living.

Homes provide a safe and familiar environment to experiment with sustainable practices and share them with family and community, fostering a sense of collective responsibility. Demonstrating how small, everyday choices can create significant positive impact inspires others to pursue a more *sustained* future.

By understanding the interconnectedness of our actions and their global impact, we can make informed decisions that benefit ourselves, the planet, and its inhabitants.

CHAPTER 2

The Conscious Kitchen: Food Waste, Climate Change, and You

We often overlook the environmental impact of our daily food choices, but the truth is the way we eat is cooking the planet. The kitchen, the heart of the home, has become a significant source of greenhouse gas emissions. Examining the entire food production process and its relation to climate change is crucial to understanding this connection better.

The food we eat and how it's processed affects our health and significantly impacts the environment. From cultivation and processing to transportation, distribution, preparation, consumption, and waste management, every step of the food production process generates greenhouse gases that trap heat from the sun, contributing to climate change. Approximately one-third of all human-caused greenhouse gas emissions can be traced back to food.

The majority of food-related greenhouse gas emissions result from agriculture and land use.

Some examples include:

- Methane produced during the digestion process of cattle.
- Nitrous oxide emissions from fertilizers used in crop production.
- Carbon dioxide released from deforestation to expand farmland.

- Other agricultural emissions from manure management, rice cultivation, crop residue burning, and farm fuel usage.

A smaller portion of greenhouse gas emissions associated with food comes from:

- Refrigeration and transportation of food.
- Industrial processes, such as producing paper and aluminum for packaging.
- Food waste management.

So, which foods contribute the most to greenhouse gas emissions? The climate impact of different foods can be compared by looking at the amount of greenhouse gases they produce when grown, processed, and transported. This is measured in a unit called "carbon dioxide equivalents," which considers all greenhouse gases, not just CO2. The more carbon dioxide equivalents a food has, the greater its impact on the climate. This measurement can be based on the weight of the food, the protein content, or the calories it provides.

Animal-based foods, particularly red meat, dairy, and farmed shrimp, generally have the highest greenhouse gas emissions.

The reasons for this include:

- Deforestation for meat production releases stored CO2 in forests.
- Cows and sheep emit methane during the digestion of grass and plants. Cattle waste on pastures and chemical fertilizers for feed crops release nitrous oxide.
- Shrimp farming often replaces carbon-absorbing mangrove forests, causing carbon release due to mangrove deforestation.

On the other hand, plant-based foods—such as fruits, vegetables, whole grains, beans, peas, nuts, and lentils—typically require less energy, land, and water and have lower greenhouse gas intensities than animal-based foods.

HOW CAN FOOD-RELATED EMISSIONS BE REDUCED?

- Promote sustainable practices from farming to consumption, lowering the food industry's carbon footprint.
- Encourage plant-based protein consumption and less animal-based food intake to decrease greenhouse gas emissions.
- Invest in plant-based substitutes, insect-based proteins, and cell-based meats to diversify options and reduce environmental impact.
- Acknowledge animal products' importance for rural livelihoods and promote sustainable livestock management.
- Use additives and techniques to lower methane emissions from cattle digestion and gases from decomposing manure.
- Manage livestock efficiently, focusing on fewer, highly productive animals to reduce emissions.
- Enhance manure and fertilizer management, implement rotational grazing for healthy soil, and restore degraded lands to lower greenhouse gas emissions.
- Minimize waste throughout the supply chain to decrease carbon footprint.
- Address food waste as a significant contributor to global greenhouse gas emissions.

Amid a global landscape that sees approximately 30 percent of food produced in the US wasted, along with the water resources utilized in its production, sustainable food systems have emerged as a beacon of hope.

The United Nations' Food & Agriculture Organization describes a sustainable food system as ensuring food and nutrition security while safeguarding the ecological, social, cultural, and economic foundations for future generations. These systems focus on conserving natural resources, embracing local and Indigenous knowledge, and respecting ecosystems, biodiversity, social equity, and human well-being. They prioritize fair, culturally acceptable, nutritious, and healthy food production across the supply chain.

Currently, 75 percent of our food supply comes from twelve plant varieties and five animal species. This reality highlights the importance of transforming our food production methods and addressing our consumption habits.

Alternative sustainable food markets and networks have recently gained popularity, allowing consumers to source food sustainably and ethically. Here are some ideas to get you started:

- **Slow Food International** advocates for countering the rise of fast-food culture by promoting good, clean, and fair food.
- **Community Supported Agriculture** (CSA) is an alternative food market that provides farmers with financial security and offers consumers a box of fresh, seasonal produce directly from the farm.
- **Solidarity purchasing groups** work together to meet a group's consumption and nutrition requirements, promoting a more localized food system.
- **Farmers' markets** offer a variety of fresh produce, meat, cheese, and bakery items from local farmers.
- **Pick-your-own farms** allow customers to pick their fruit, vegetables, and more, reducing harvesting labor costs for farmers.
- **Box schemes** provide consumers with various fruits and vegetables to encourage diverse consumption.
- **Upcycled Food Enterprises** minimize food waste by selling oddly-sized produce or upcycling it into processed food.
- **Direct trade** bypasses middlemen and sources food directly from the producer, supporting fair prices for farmers.
- **Co-ops** offer locally sourced produce, meat, dairy, and pantry staples, allowing consumers to invest in local food systems.

TIPS TO SUPPORT A MORE SUSTAINABLE FOOD SYSTEM

- Eat organic food (if you can afford it)
- Support local farmers and small businesses by purchasing their produce
- Improve cooking skills with fresher ingredients
- Choose in-season fruits and vegetables
- Promote fair and direct-trade practices
- Endorse eco-friendly farming methods: organic, biodynamic, or regenerative agriculture
- Reduce food waste and use eco-friendly packaging
- Look for alternative food sources and environmentally-conscious brands
- Start personal or community garden plot
- Collect and distribute seeds for future harvests
- Eat less meat
- Share culinary knowledge: recipes, methods, nutrition, and local customs

While burning fossil fuels for electricity, heat, and transportation is the largest source of greenhouse gas emissions in the United States, food waste in our kitchens is closely linked to methane emissions, which are even worse for the environment than CO_2.

In 2021, US landfills released 122.6 million metric tons of methane, accounting for 16.9 percent of the country's total human-caused methane emissions.

Given these alarming figures, tackling the pressing issue of reducing kitchen waste is crucial. The world wastes 1.4 billion tons of food annually, but the US alone discards 103 million tons (206 billion pounds), surpassing any other country. This wasted food could have provided almost 130 billion meals. This is due to overbuying, poor planning, confusion about expiration dates, inadequate storage, and large portions. Excessive food purchases result in spoilage and waste, while non-food waste, including plastic packaging and disposable items, only compounds the problem. It's time to turn the heat down on waste and reevaluate our kitchen habits for a more sustainable future.

Conducting a DIY waste audit is crucial before cooking up waste reduction strategies in your kitchen. This audit uncovers your current waste generation habits, helps set realistic waste reduction goals, tracks progress over time, and motivates continuous improvement. Moreover, waste audits can serve as a delicious helping of financial savings by reducing the amount spent on groceries that eventually go to waste. In essence, reducing waste is not only eco-friendly but also eco-nomical!

To conduct an at-home waste audit, follow these steps:

1. **Choose a timeline:** Pick a typical week with no special events or activities, like holidays or parties, to monitor your waste output for seven days. This helps to get an accurate picture of your everyday waste habits.

2. **Collect your trash:** During the chosen week, gather all the waste generated in your kitchen. You could use separate bins or bags for different waste categories like glass, plastic, paper, and food. This makes the audit process cleaner and more organized.

3. **Set up your audit area:** After your week of collecting garbage, find a spacious, easy-to-clean area (ideally indoors) to sort your waste. Cover the surface with a tarp or protective material to make cleanup easier.

4. **Sort your waste:** Divide the trash into four categories: curbside recyclables (plastic bottles, glass, metal, paper), household hazardous (Styrofoam, batteries, paint), organic waste (food scraps, compostable materials), and everything else (e-waste, textiles).

5. **Record your results:** List each item in every category and note the quantities. This data will give you a clear picture of your household's weekly waste production and the types of waste generated (see chart below).

6. **Create a waste reduction plan:** Based on your findings, identify areas where you can significantly reduce waste. Consider strategies like meal planning, proper food storage, or finding ways to repurpose or recycle specific items.

Remember to repeat the waste audit twice yearly to track your progress and adjust your waste reduction plan as needed. I cannot stress the importance of knowing your area's recycling rules! Properly sorting and disposing of recyclable items makes recycling smoother and more efficient. In doing so, we're helping Mother Nature, conserving precious resources, and minimizing our impact on the planet.

	REUSE	RECYCLE	LANDFILL	COMPOST	OTHER	ALT.
cardboard egg carton		✓		✓		
plastic produce bag			✓			cloth bag
paper		✓		✓		
batteries					✓	
textiles	✓		✓			clothing swap or donate
dryer sheets			✓			wool dryer balls
food scraps				✓		make your own
nut milk container		✓				
glass jars	✓	✓				buy in bulk
metal cans		✓				

My waste audit was an eye-opener! Shocked by the recyclable or repurposable items in my trash, I realized I could make better choices. My new mission was to swap single-use plastics for reusable glass and choose packaging-free options. I embraced bulk shopping and farmers' markets, picked fresh produce over packaged, and switched to cardboard egg cartons and glass dairy containers. This trash transformation helped me join the circular economy and adopt a more sustainable lifestyle.

With the waste audit complete, let's dive into the next step: understanding the difference between "sell by," "use by," and "best by" dates. Knowing these distinctions will help reduce food waste, enhance sustainability, and save money. "Best by" dates are not expiration dates; they simply indicate the last day before a food product's quality starts to decline. Milk, for example, can still be good for up to a week after its printed date.

Food dating emerged in the 1970s when consumers sought information about their food's production. As a result, companies implemented date ranges to maintain food quality and consumer satisfaction. Today, "best by" indicates when a product will have the best flavor or quality, while "sell by" ensures proper turnover during a food product's journey, helping it maintain a long shelf life after purchase. "Use by" marks the last date recommended for the product's consumption, often found on meat, poultry, or egg labels, and should be taken seriously. Eating food past its "best by" date may only result in a decline in quality, but consuming items past its "use by" date can cause illness, with symptoms ranging from a stomach ache to severe food poisoning. To determine if food is safe after its expiration date, differentiate between "best by" and "use by" dates, considering proper storage and unopened packaging.

The Food and Drug Administration (FDA) does not have formal regulations for food dating. The Food Marketing Institute and the Grocery Manufacturers Association established their rules, which many food companies have adopted. "Best by" and "sell by" indicate quality, while "use by" indicates safety. Despite these guidelines, always exercise common sense when consuming products past their dates. I like to follow my nose. If it smells funky, it's probably time to go.

FIGHTING FOOD WASTE

Approximately 30 to 40 percent of the US food supply is wasted at 219 pounds per person. It's like every American tossing over 650 average-sized apples straight into the landfill, where most discarded food ends. Food comprises 22 percent of municipal solid waste in US landfills. Wasting food has dire environmental consequences; it squanders the water and energy used in production and generates greenhouse gases, such as methane, carbon dioxide, and chlorofluorocarbons, contributing to global warming. Decaying food in landfills also produces nitrogen pollution, leading to algae blooms and dead zones. The World Wildlife Federation states that the production of wasted food in the US is equivalent to the greenhouse emissions of thirty-seven million cars. This is why keeping food out of the garbage can is crucial.

> **FACT:** The average American family of four throws out $1,600 a year in produce.

Given that America wastes about one pound of food per person daily, it's crucial to recognize that proper food storage plays a key role in reducing food waste. One way is by setting up a **capsule pantry**. Like a capsule wardrobe, a capsule pantry comprises essential items that work together and stand the test of time. Susie Faux, a London boutique owner in the 1970s, coined the concept. A capsule pantry simplifies your approach to food and includes ingredients you love and use to create multiple meals. It's important to note that this concept applies to dry goods and the fridge and freezer.

Here's a step-by-step guide to help you establish a successful capsule pantry and minimize food waste:

1. **Assess your current pantry:** Take inventory of the items you already have. List the staples you use most frequently and those sitting around too long. This will give you an idea of what you truly need

and use. Compost anything that's expired. If they are unopened and nonperishable, donate them.

2. **Determine your essentials:** Create a list of pantry essentials you can use in multiple dishes based on your cooking habits and preferences. These may include grains, legumes, canned goods, spices, oils, and vinegar.

3. **Plan for variety:** To avoid boredom and promote creativity in the kitchen, ensure your capsule pantry includes ingredients with diverse flavors, textures, and uses. This will encourage you to experiment with different dishes and minimize the temptation to eat out or waste food.

4. **Streamline your shopping:** Armed with your list of essentials, shop mindfully and avoid impulse purchases. Stick to your list, and consider buying in bulk when it makes sense for items you use frequently. This not only helps reduce packaging waste but can also save you money.

5. **Organize and store properly:** To maintain a successful capsule pantry, keep it organized and store items to maximize its shelf life. Use clear, airtight containers to store dry goods and label them with expiration dates. Rotate items, placing the oldest products in front to ensure they're used before newer ones.

6. **Meal plan and prep:** Plan your meals around the ingredients in your capsule pantry. This will help you utilize your items and reduce the likelihood of food waste. Prepare some ingredients in advance, such as prepping veggies or cooking grains, to make mealtime more efficient and enjoyable.

> **PRO TIP:** Save money on groceries by creating a rotating meal list featuring frequently used ingredients. Stock up on staple pantry items like flour and oil when they're on sale, saving hundreds of dollars. Organize your pantry by placing items with the earliest expiration dates at the front to ensure they're used first. This smart shopping and storage strategy helps you prepare your favorite meals without breaking the bank.

ZERO WASTE PANTRY CHECKLIST

GRAINS
- ☐ rice
- ☐ oats
- ☐ pasta
- ☐ quinoa
- ☐ noodles
- ☐ cereal

LEGUMES
- ☐ beans
- ☐ lentils
- ☐ chickpeas

NUTS & SEEDS
- ☐ almonds
- ☐ cashews
- ☐ walnuts
- ☐ pecans
- ☐ sesame seeds
- ☐ chia seeds
- ☐ pumpkin seeds

SPICES & DRIED HERBS
- ☐ oregano
- ☐ basil
- ☐ cumin
- ☐ garlic powder
- ☐ salt + pepper
- ☐ onion powder
- ☐ nutritional yeast
- ☐ tumeric

SAUCES, OILS, & VINEGAR
- ☐ olive oil
- ☐ sesame oil
- ☐ white vinegar
- ☐ apple cider vinegar
- ☐ balsamic vinegar
- ☐ tamari
- ☐ soy sauce

BAKING
- ☐ baking soda
- ☐ cocoa
- ☐ chocolate chips
- ☐ flour
- ☐ sugar
- ☐ cornmeal

LESS WASTE, MORE COMPOST

According to the UN, the world's urban population is predicted to increase to 68 percent by 2050, resulting in approximately 2.5 billion people living in condensed urban spaces. This growth in urban areas has led to increased waste production, with the World Bank's *What A Waste 2.0 Report* finding that urban dwellers generated 2.01 billion tons of waste. This amount is expected to rise by 70 percent to 3.4 billion tons by 2050. Shockingly, an EPA study found that 56 percent of nonindustrial food waste in the USA ended up in landfills, contributing to methane and other greenhouse gas emissions. Only 4 percent of the food waste was composted, highlighting the importance of learning how to compost to help reduce waste and promote a more sustainable future.

Composting transforms kitchen scraps and yard waste into a natural fertilizer that promotes healthy plant growth, reducing the need for chemical fertilizers that can harm the environment. This process not only diverts organic materials from landfills but also mitigates methane emissions. Additionally, composting saves money by eliminating the need for commercial fertilizers while fostering a deeper connection to nature and encouraging a sustainable lifestyle through mindful waste management and repurposing.

There are two standard methods to compost at home: utilizing a city composting program or setting up a compost bin in your yard or garden.

1. **City Composting Program:** Many cities offer composting programs that provide designated bins or containers for organic waste. Residents can collect their food scraps and yard waste in these containers, which are then picked up by the city's waste management service regularly. The collected waste is taken to a central composting facility and processed into compost. This method is ideal for those needing more space, time, or resources to maintain a compost bin in their yard.

2. **Home Composting with a Compost Bin:** Setting up a compost bin in your yard or garden allows you to manage the composting process on your property. Various compost bins are available,

from simple DIY setups to more advanced, commercially available options. The composting process involves adding a mixture of green materials (e.g., food scraps, grass clippings) and brown materials (e.g., dried leaves, small branches) to the bin and occasionally turning the pile to aerate it and promote decomposition. Over time, the organic waste breaks down into nutrient-rich compost that can be used in your garden or yard. This method requires more hands-on involvement but provides a direct source of plant compost.

How to Compost in an Apartment

Apartment buildings frequently handle their waste management independently, which might result in the absence of organic waste collection services. This happened to my husband and me for years in Toronto. It was frustrating; we reached out to the building owner many times to try and rectify this, but unfortunately, it fell on deaf ears. We ended up setting up vermicompost. (More on that in a bit.) Many people believe that composting is only possible with access to a backyard or outdoor space, and the potential odor may deter apartment residents from trying it. However, when done correctly, you can compost organic waste in your apartment or balcony without attracting pests or creating unpleasant smells. If you're living in a smaller space and have contemplated starting a compost bin but feel uncertain, let this serve as an inspiration and a beginner's guide to apartment composting.

Decide on Your Composting Method

The first step to apartment composting is deciding which composting you want to try.

TRADITIONAL COMPOSTERS

The Envirocycle Mini Composter is a compact, all-in-one system for small spaces like balconies or patios. This seventeen-gallon, tumbler-style composter is efficient, easy to use, and aesthetically pleasing. It produces both solid compost and nutrient-rich liquid compost tea. Utilizing decomposition (aerobic) or fermentation (anaerobic) processes to break down food scraps, the Envirocycle Mini is better suited for balcony composting (outdoors) due to the natural odors emitted during these processes.

VERMICOMPOSTING

Vermicomposting, which can be used indoors or outdoors, is an excellent method for processing organic waste in apartment settings. Worm bins, such as Sacred Resources Composter and Living Composter, utilize red wriggler worms to break down organic materials, producing nutrient-rich castings and worm tea that gardens thrive on. Worms are an odorless and eco-friendly alternative for composting in apartments or balconies. However, it's essential to remember that worm farms involve living creatures and may require more attention. Ideally, worm farms are suited for apartments with balconies, but a sturdy bin can facilitate indoor vermicomposting. Therefore, taking care of your worm farm is crucial.

Here are some essential tips:

- Worms need a moist environment and should be kept away from excessive heat. Pour a large bucket of water through the bin at least once a week. The worm tea produced can be drained and used as an excellent fertilizer for indoor and outdoor plants.
- Provide food at least once a week, but avoid overfeeding to ensure they have enough space to move around.
- Worms have specific dietary restrictions. Refrain from feeding them onions, citrus, and other items (refer to a comprehensive list for details).
- Routinely monitor and clean the worm bin to ensure efficiency and effectiveness.

When I initially tried worm composting, I was apprehensive due to the "ick factor." However, my discomfort subsided as I learned about the process and its benefits. Understanding the science behind vermicomposting and maintaining a clean system helped minimize unpleasant smells. Using gloves and tongs for handling made the process more comfortable, and eventually, I began to appreciate the worms as eco-friendly helpers contributing to a sustainable lifestyle.

BOKASHI BINS

Bokashi bins, suitable for indoor or outdoor use, offer another composting option for apartment dwellers. This Japanese technique involves "pre-composting" food scraps through fermentation with the help of microorganisms. Bokashi bins can handle a broader range of waste, including items unsuitable for worms, such as meat, citrus, and other acidic foods. Bokashi bins complement vermicomposting systems since the end product requires further decomposition by burying in soil or feeding worms. These bins are a viable choice for apartment composting, even without worms, and are available globally. Some options include Bokashi Living and Sunwood Life Bokashi. These are ideal for indoors or outdoors.

HIGH-TECH COMPOSTERS

The Vitamix FoodCycler is an electric composter that reduces food waste volume by up to 90 percent. It combines heat, grinding, and aeration to break food scraps into a nutrient-rich soil amendment. The process takes only a few hours and is virtually odor-free. The end product can be used in gardens or with potting soil for indoor plants. The Lomi is another high-tech kitchen appliance that uses heat, abrasion, and oxygen to break down organic waste. It can process various food scraps, including fruit peels, vegetable trimmings, and even small bones. Like the FoodCycler, the Lomi reduces waste volume significantly and produces nutrient-rich soil in just a few hours. Both are meant to be used inside.

I've used both machines, and despite these similarities, they differ in price, size, noise, capacity, and materials. The FoodCycler is generally more expensive but has a larger capacity and operates more quietly. On the other hand, the Lomi is more compact and made from eco-friendly materials. Ultimately, both devices offer practical and sustainable solutions for managing organic waste in small living spaces.

Composting is essential to sustainable living; it's been a fulfilling and rewarding activity for me. I love composting because it reduces waste, conserves resources, and improves soil health. Composting gives me a sense of connection with the natural world and allows me to participate in the cycle of life by turning waste into a valuable resource.

MASTERING FOOD STORAGE WITH THE RIGHT TOOLS

Proper food storage is critical in preventing food waste and ensuring that the food you bring into your home stays fresh and safe for consumption. When we store our food correctly, we can extend its shelf life, maintain its nutritional value, and preserve its flavor and texture. This helps us save money by reducing the need to purchase replacements and contributes to a more sustainable lifestyle by minimizing the environmental impact of food waste.

First, don't overcrowd your fridge. This can lead to restricted airflow, causing your fridge to work harder, consume more energy, and create "warm spots" where food may spoil faster. Additionally, when your fridge is too full, it's harder to see what you have, leading to forgotten items that eventually go bad. Another critical step in proper food storage is separating "gassy" fruits and vegetables. Some produce, like ripe bananas, avocados, cantaloupe, potatoes, and pears, emit ethylene gas as they age. This gas can cause more sensitive fruits and vegetables, such as leafy greens, onions, cucumbers, and carrots, to age more rapidly. So, keeping these "gassy" items separate from the more sensitive ones is vital. Investing in airtight containers can also make a significant difference in reducing food waste. For example, if your family tends to go through cereal, crackers, cookies, nuts, or other dried snacks quickly, you don't need to transfer them to airtight containers. However, if these items frequently go stale, investing in airtight containers can save you money in the long run. For this purpose, you can even repurpose glass or plastic jars from other food items, like pasta sauce or peanut butter. Just make sure to clean and dry them thoroughly before use. When it comes to freezing food, it's essential to consider how you'll use it later. For example, while it might be tempting to freeze meat directly in its packaging or store a large portion of lasagna in one container, this approach can lead to inconvenient and wasteful outcomes. Instead, trim, cut, or portion meat into sizes you'll use when cooking, and divide leftovers into individual servings before freezing. This way, you'll be more likely to use the frozen items instead of letting them sit in the freezer until they're no longer suitable. Refrain from assuming that the fridge is always the best place for food storage. Some foods spoil or become stale faster when stored at cold temperatures. For example, melons should be stored at room temperature until cut, while bread goes stale more quickly in the fridge. Cold temperatures can also negatively affect the texture of tomatoes and potatoes and diminish the quality of coffee beans. So, it's essential to research the best storage conditions for different foods to ensure they stay fresh as long as possible. Lastly, when it comes to leftovers, store them front and center in your fridge and at eye level. This simple trick will help you remember to use them quickly and prevent them from ending up in the trash. Leftovers are a significant source of household food waste, so being mindful of their storage can make a big difference.

Reusable Silicone Bags

Although technically a type of plastic, reusable silicone bags are a far superior alternative to plastic wrap and freezer bags. They are designed for reuse and are safer than even the more durable plastic food containers. In addition, they provide the same benefit of removing air from the bag to preserve freshness.

Reusable silicone bags last longer and are more effective at reducing air and moisture transmission due to their thickness and durability. I wish they were biodegradable and had better recycling rates.

Tempered Glass Containers with Airtight Lids

Tempered glass is a popular choice for plastic-free storage containers and works well in the freezer. Ensure you select a tempered glass or freezer-safe option and choose a container with an airtight lid to create the tight seal needed to protect the contents from air and odors in the freezer. Remember not to fill glass containers to the brim, leaving some space at the top to prevent broken glass in your freezer. With these precautions, you have a simple, plastic-free option that may already be in your home.

Stainless Steel

These containers with airtight lids are also an excellent choice for storing meat in the freezer without plastic. These containers are perfect for both leftovers and raw meat.

Compostable Parchment Paper

For added protection, compostable parchment paper is a great option. Compostable parchment paper is biodegradable and can be added to your home compost heap. Then, when you're ready to compost it, cut it into small pieces, just like you would with veggie scraps and other compostable waste.

> **PRO TIP:** To freeze meat, wrap it in compostable parchment paper before placing it in a tempered glass or stainless steel container. This method may be best for more manageable cuts of meat, as more oversized items like whole chickens could be more challenging to handle.

Beeswax Wraps

Beeswax wraps are an eco-friendly alternative to plastic wrap, made from fabric infused with beeswax, jojoba oil, and sometimes resin. They are washable, reusable, and compostable. To use beeswax wraps, mold them around the food or container you want to cover using the warmth of your hands. The beeswax will soften slightly and adhere to itself, creating a seal that keeps food fresh. They are great for wrapping fruits, vegetables, cheese, and sandwiches or covering bowls and containers.

To clean beeswax wraps, gently wash them in cool water with mild dish soap and then air dry. Avoid using hot water or exposing the wraps to heat, as it can melt the wax and damage the wrap. In addition, they should not be used for wrapping raw meat or fish due to the inability to sterilize them in hot water.

If your beeswax wrap is holding onto a smell, like blue cheese or onion, let it air dry for a few days. The smell should dissipate. You can also consider using specific wraps for certain smelly foods. Store away from heat (stove) or humidity (dishwasher) in a cool, dry place. Keep some folded and stored in a drawer with your reusable paper towels and a few gently rolled up on the kitchen countertop for easy access.

I wanted to write this book to help you be more eco-nomical, which means dispelling the myth that living more sustainably is more expensive—it's not. Many things in our homes can be repurposed or used creatively to reduce waste and save money. Sustainable living is not about buying the latest eco-friendly gadgets or products but about making conscious choices and taking small steps toward a more sustainable lifestyle.

ECO-NOMICAL SAVINGS—BEESWAX WRAPS VS. PLASTIC BAGGIES

A pack of fifty medium Ziploc bags costs around $8.35, about $0.17 per bag. Using five Ziploc bags per week will cost around $40.80 per year on Ziplocs. On the other hand, one beeswax wrap typically costs around $12, but they can last up to a year with proper care and use. Assuming you replace your beeswax wraps yearly, you would save over $28.80 annually using beeswax wraps instead of Ziploc bags.

ECO-NOMICAL SAVINGS—REUSED GLASS JARS VS. PLASTIC FOOD STORAGE

A set of ten Rubbermaid Brilliance containers with lids costs around $22.99 on Amazon, which comes out to $2.29 per container. On the other hand, reusing glass jars is virtually free, as they are commonly found in households and can be saved from items like spaghetti sauce or pickles. So, by reusing glass jars instead of buying new plastic containers, you can save up to $2.29 per container. Over time, this can add up to significant savings!

FROM MARKET TO MEALTIME

Consider buying the freshest products from your local butcher or directly at the meat counter. Bring your own stainless steel or glass container, ask for the tare, and have your meat placed directly in the container without plastic trays, wrap, or butcher paper. Plan to freeze meat immediately after purchasing. Avoid unnecessary stops in the refrigerator. First, decide which meat will go in which container. Then, consider portioning meat like ground beef or chicken breasts into servings that work best for your household or meal plans. Prepare your containers, find something to write the date to, and label your containers—clear space in your freezer before handling the meat. Aim for the coldest part of the freezer, usually the bottom shelves or drawers.

Prepare and Label Your Meat for the Freezer

If using silicone bags, remove as much air as possible by placing the bag with the meat on the counter, partially closing the zipper, guiding the air out with your hands, and then closing it completely. Don't worry about getting every air bubble out—do your best. Place the meat directly in the container for stainless steel or tempered glass containers (remember to leave room in glass containers) and ensure a proper seal when fastening the lid. Always label items before placing them in your freezer. For example, you may recall that a jar contains marinara sauce, but it may be challenging to differentiate between plain tomato and marinara after a couple of months.

If possible, place them in the coldest part of the freezer to maintain freshness and prevent freezer burn. Consider using a thermostat to ensure the temperature remains at 0°F (-17.78°C) or below. If you plan to maintain a constant supply of frozen meat, remember to rotate the contents in the back of your freezer to the front and restock as needed. While frozen meat remains safe to eat indefinitely when kept at a temperature of 0°F (-17.78°C) or lower, its quality will degrade over time, even with the best freezing techniques.

When it comes to freezing meat, I'm a big advocate for avoiding the use of plastic. It's common for people to freeze meat in airtight or vacuum-sealed disposable plastic bags, while others choose cellophane. Although these materials effectively protect the meat from freezer burn, they significantly negatively impact the environment and our health. If you're thinking of reusing disposable freezer bags, it's not advisable, as washing them can lead to degradation and potential contamination. When freezing food, the goal is to minimize air and moisture exposure to prevent freezer burn.

> **OTHER WAYS TO PRESERVE FOOD**
>
> **Pickling:** This process, which has been around since ancient times, involves immersing food items in an acidic solution, typically vinegar. It's a foolproof way to preserve everything from cucumbers to carrots, radishes, and avocados.
>
> **Fermenting:** Fermenting can be tricky as it involves creating the right environment for bacteria to react with the sugars in food. But once you get the hang of it, you can make all sorts of tasty treats like sourdough, kimchi, and kombucha.
>
> **Dehydrating:** Dehydrating extends the life of various food items, from herbs and spices to fruits. Whether using an oven, the sun, or a dehydrator, it's a simple and effective way to minimize food waste.
>
> **Canning:** Canning is more involved, requiring the right tools and safety precautions. But when done correctly, it can keep food fresh for years. That's certainly worth the effort.
>
> **Juicing:** Got a juicer or a blender? Then you have a quick and easy way to save fruits and vegetables from going to waste. You can enjoy these juices immediately or add them to smoothies, soups, and more.
>
> **Curing:** The National Center for Home Food Preservation defines curing as the addition of a combination of salt, sugar, and nitrite to meats. To start curing, check out the plethora of resources available online.

Get to Know the Bulk Section

Bulk options are increasingly available in grocery chains and health food stores. Use your DIY waste audit to find bulk alternatives, and if not available, opt for items in glass, aluminum, or cardboard. Use online recipes to make items like hummus, guacamole, or nut milk. Choose containers marked 1 or 2 for higher recycling rates if plastic is necessary. Buy larger containers instead of individual servings to reduce waste. Sticking to the produce section helps reduce waste and simplifies life. Buying less saves money, promotes healthier eating, and reduces food waste. Bulk buying allows you to purchase only what you need, minimizing waste. For example, buying oatmeal from bulk bins saves five times the waste compared to packaged options.

Shop at a Local Butcher or Baker

Shopping at your local butcher or baker can help reduce waste by allowing you to buy just what you need rather than prepackaged items. In addition, your local butcher or baker will likely give you much less resistance when bringing your containers. It is easier to pay for your items without the sticker we usually need for the cash register in larger shops. You will also reduce your food miles, support your local economy, and nurture a sense of community. You bring your containers and place one on the scale for the tare, after which your items will be added into the jar and weighed for pricing. Remember to ask if you can be checked out without the sticker.

Plan, Plan, Plan

Have a selection of go-to meals and prioritize easy, healthy recipes. Before shopping, check your fridge and freezer, then list what you need. Stick to it!

Maintain a well-stocked pantry and freezer with items that don't have short "best before" dates. Shop in bulk for rice, pasta, and quinoa, with frozen fruits, veggies, fish, and chicken on hand. This allows you to prepare healthy meals quickly.

Repurpose leftovers, like turning old tomatoes into pasta sauce. While fresh food is often better, shopping seasonally and freezing produce is a mindful way to eat. When buying in bulk, make sure you'll use the items before they expire.

A word of advice about buying bulk at a place like Costco, make sure you know you will get to it. We tend to buy more stuff we do not need at shops like this, so be mindful!

Portion Control

I'm not talking about weight loss here. Cook enough food, perhaps even extra for leftovers, and utilize a food portion calculator. Remember to read the cooking instructions carefully as well. Put those leftovers to good use. If you roast a chicken, keep the water at the bottom as chicken stock, make a yummy chicken and mayo sandwich the next day, and use the carcass to make a fantastic bone broth. Think outside the box and Google "leftovers." There are so many great resources online to help you.

Buy Ugly Food

Ugly food, also known as "imperfect produce," refers to fruits and vegetables considered too "ugly" or imperfect to be sold in traditional grocery stores. This can include items with blemishes, misshapen produce, or items that are smaller or larger than average. While these items are perfectly edible and nutritious, they are often discarded and go to waste because they need to meet the aesthetic standards of many retailers and consumers. Fortunately, the movement to reduce food waste has increased the availability of ugly food. Many retailers and organizations are now selling or distributing imperfect produce at a lower cost, making healthy food accessible to a wider range of people while reducing waste. Additionally, buying ugly food can help support sustainable agriculture practices and reduce greenhouse gas emissions associated with food waste. We can achieve a more sustainable and equitable food system by embracing ugly food.

MY FAVORITE SAVING-FOOD HACKS

- Use up that last bit of yogurt for smoothies or cereal toppings.
- Steep fruit peels, like apple and orange, in hot water for a simple homemade tea. Stir in a little honey for a delicious, warm drink.
- Don't store your eggs on the fridge door because it's the warmest part of the fridge. Store it on one of the middle or top shelves instead.

- If you buy discounted bread near its sell-by date, freeze it and only thaw as many slices as you need at a time.
- Grate lemon, lime, or orange peels to use as zest in recipes or infuse them in water or vinegar for natural cleaning solutions.
- Start a "garbage soup" container in your freezer. Collect leftover meat and vegetable scraps; cook them in a pot with broth when it's full. Tailor the seasoning based on the contents.
- Use strawberry tops to create an infused vinegar.
- Repurpose pickle brine by adding peeled carrots. After a week, they develop a delicious pickled taste, ideal for a relish tray or a Bloody Mary.

REDUCE, REUSE, SHOP! HOW TO USE REUSABLES

My dad always said, "Prior planning prevents poor performance." I believe in having the right tools for the job; the following list is my go-to. You do not have to go out and purchase all of these new items. First, shop your home. I bet you have many of these on hand already. I've found some of my favorite reusables at my local thrift shop for a glass mason jar for as little as $1.99. I've also created a buy-nothing group in my neighborhood. You can do this too! These groups are growing in popularity, super-localized, and no money is exchanged; you can shop, borrow, or swap. Hello, sharing economy!

Reusable Bags

Essential to a sustainable lifestyle, reusable bags from materials like cotton, canvas, or nylon come in various sizes. Keep them near your front door, car, or glove compartment for easy access. A well-rounded collection should include the following:

- Four to five thick grocery bags.

- Two insulated bags for cold items.
- Up to five fabric or cotton totes for lighter purchases.

> **PRO TIP:** Include "bags" at the top of your shopping list as a reminder to bring them with you.

USE YOUR BAGS PROPERLY

- Empty bags entirely after use.
- Wash all bags regularly.
- Use bags that are easy for the cashier to fill.
- Open bags that fold up into themselves while you are waiting in line. Don't make the cashier wait for you to open them.

TYPES OF BAGS

- **Compact:** Small enough to shove in your pocket.
- **Comfortable:** You can sling it over your shoulder with a long handle.
- **Self-contained:** Rolls up into itself or a built-in little holder bag.
- **Big and strong:** Can carry a heavy load of library books or groceries.
- **Free bags:** Don't take free bags at events! You know, the ones they hand out for free crap we don't need. They usually are cheaply made and, in many cases, contain plastic (polyester).

Mesh bags simplify eco-friendly shopping for produce and bulk items. Keep four on hand to ease the checkout process and double as storage and washing bags at home. In the produce section, use mesh bags for heavier items like yams, potatoes, and apples, and opt for muslin bags for leafy greens and mushrooms. Utilize glass jars at deli counters for olives, cheese, and meats, and choose bagels from bins instead of plastic packaging. Carry a foldable grocery bag in your backpack or purse for unexpected shopping trips, and keep extras in your trunk. Ask the store staff for empty boxes or paper bags if you need a bag. When faced with resistance using your containers, remember your power as a customer. Engage in polite conversations with shop owners or managers to encourage change. Stay polite, clear, and understanding throughout the process. The more we advocate for sustainable practices, the more impact we'll have.

Repurposed Glass Jars and Metal Containers

Metal and glass containers are ideal for buying meat and cheese due to their ability to maintain freshness and hygiene. They are lightweight, durable, easy to clean, reusable, and free of harmful chemicals. Start with three to five versatile containers in different sizes to accommodate your needs, and adjust the quantity as needed. Don't discard old plastic containers; repurpose them for storage or other household uses, such as organizing supplies or making an inexpensive compost bin. Use them for bulk shopping dry foods, transferring the contents to airtight containers at home.

HOW TO SHOP FOR NONTOXIC COOKWARE

Nonstick options have been popular for decades due to their ability to repel oil and water. Still, many of these products contain harmful chemicals such as perfluoroalkyl and poly-fluoroalkyl (PFAS). PFAS is a group of man-made chemicals used in various industries since the 1940s, including in producing nonstick cookware, stain-resistant fabrics, water-resistant clothing, food packaging, and firefighting foams. PFAS are considered "forever chemicals" because they do not break down easily and can persist in the environment and our bodies for hundreds, if not thousands, of years. As a result, they have been linked to numerous health issues, including cancer, liver damage, thyroid disease, and developmental issues in fetuses and breastfed babies of exposed women.

PFAS can be found in a variety of products, including nonstick pans, food packaging (like microwave popcorn bags and pizza boxes), waterproof and stain-resistant products (like jackets and carpets), and personal care items like shampoo, hairspray, mascara, and nail polish. These chemicals can enter the body through inhalation, ingestion, or skin absorption. Additionally, PFAS have been found in drinking water near manufacturing or usage facilities, such as military installations, wastewater treatment plants, and firefighter training facilities. The CDC reports that PFAS have even been detected in remote locations like the Arctic and the open ocean.

Because the phase-out of PFOA and PFOS has been voluntary, some companies may still use these chemicals in food contact materials. This highlights the importance of being informed about the products we buy and the materials they are made of and advocating for stricter regulations to protect our health and the environment. PFOA and PFOS have also been added to California's Proposition 65 list of chemicals that cause cancer or reproductive harm. California recently became the first state to pass legislation to phase out PFAS from clothing and textiles, with the European Union considering a similar ban. To delve deeper into the issue, I recommend watching the movie *Dark Waters*, the documentary *The Devil We Know*, or reading *Exposure* by Robert Bilott. The latter two are based on a true story that reveals the scandalous history of the poisoning of a West Virginia town by the makers of Teflon. Although fascinating, it is also angering to see how corporations prioritize profits over human health and the environment.

Greenwashing is a common problem in the nonstick cookware industry. Many brands claim "PFOA-free" to attract health-conscious consumers, but this does not necessarily mean the product is free of all PFAS. As the harmful effects of PFOS and PFOA became known and were phased out in the US, they were replaced with other chemicals from the PFAS family, such as PFHxS, PFNA, and GenX. Although initially believed to be safer, we now know they are not. Nonstick cookware labeled as "PFOA-free" and "PFOS-free" often contain polytetrafluoroethylene (PTFE), a different type of polyfluoroalkyl substance in the same family as PFAS chemicals.

In the introduction to *Sustained*, I discuss the importance of knowing how to shop for certain sustainable products using my three criteria. These can be applied here:

Criteria 1: What Is the Item/ Product Made From?

Look for cookware free of *all* PFAS, including PTFE. Also, check for cookware tested and verified against Prop 65 guidelines. This is a way to identify brands tested for heavy metals such as lead, cadmium, and certain PFAS. Choose brands that provide transparency and third-party testing from certifications like Registration, Evaluation, Authorization, and Restriction of Chemicals (REACH), and Proposition 65. Ideally, look for brands that disclose ingredient information and provide testing to support their claims. However, obtaining the most thorough and up-to-date information may only sometimes be possible. In such cases, make the best decision based on the information available.

Criteria 2: Sustainable Sourcing/ Ethical Manufacturing

The production of cookware requires the extraction of metals and minerals through mining, which has been linked to exploitative and harmful practices. Therefore, ethical sourcing and manufacturing are crucial in this industry. However, even the most sustainable brands have yet to achieve complete transparency in their supply chains, and relying solely on certifications may not ensure sustainability. Nevertheless, some brands are taking steps to promote fair wages, safe working conditions, and gender equality, which are validated by certifications such as Business Social Compliance Initiative (BSCI), Sedex Members

Ethical Trade Audit (SMETA), Supplier Ethical Data Exchange (SEDEX), Social Accountability International (SA8000), ISO 9001, ISO 14001, and ISO 45001.

Criteria 3: Corporate Social Responsibility

Seek brands demonstrating their commitment to environmental and social responsibility. These might include warranties for their products, carbon-neutral shopping, and eco-friendly packaging materials.

BEST NONTOXIC COOKWARE & BAKEWARE

Over time, cookware can become worn out, scratched, or damaged, making it unsafe. For example, nonstick coatings on pots and pans can start to break down, releasing harmful chemicals into your food. The same is true for other types of cookware, like aluminum, which can leach into your food and potentially cause health issues. When it comes to worn-out cookware, it's often best to replace it with new, safer options, especially if it's showing signs of wear and tear.

Cast Iron

For several reasons, cast iron is popular for nontoxic, sustainable cookware. It's made from a natural material with no harmful chemicals or coatings that can leach into your food. It is also highly durable and can last for generations with proper care. In addition, cast iron cookware is versatile and can be used on all stovetops, including induction and even in the oven. Another advantage of cast iron is that it retains heat well, making it great for searing meats and cooking at high temperatures. It also has natural nonstick properties that improve with seasoning, which builds up a layer of oil on the pan's surface over time. Seasoning enhances the nonstick properties of cast iron and protects the cookware from rust and corrosion. However, there are some cons to using cast iron cookware. Cast iron is heavy, making it difficult to handle for some people, and it can take a while to heat. It can also be prone to rust if not appropriately seasoned or left wet. Additionally, acidic foods like tomatoes and vinegar can react with iron and give the food a metallic taste. Finally, cast iron cookware requires more maintenance than other cookware and cannot be cleaned in the dishwasher.

Brand recommendations: Lodge 💡, Field Company, Nest Homeware

Ceramic

Ceramic cookware comes in two types—100 percent ceramic and ceramic-coated. 100 percent ceramic cookware is made from sand, clay, quartz, and other minerals. It is eco-friendly, containing no harmful chemicals such as PFAS, PFOA, or heavy metals. It is also safe to use as it does not release toxic fumes when heated. The benefits of 100 percent ceramic cookware include even heat distribution, durability, and versatility, as it can be used on various cooktops. However, it is heavy, expensive, and can break if dropped.

Brand recommendations: Xtrema, Emile Henry

Ceramic-Coated Cookware and Sol-Gel

Ceramic-coated cookware, on the other hand, has a metal core wrapped in a ceramic outer layer. Although it may contain harmful chemicals such as PFAS or PFOA, it is still safer than traditional nonstick cookware. Ceramic-coated cookware has advantages such as being nonstick, easy to clean, and heat quickly and evenly. It is also less expensive than 100 percent ceramic. However, it is less durable than 100 percent ceramic cookware, and the ceramic coating can wear off over time. In addition, ceramic-coated cookware is less versatile and cannot be used on high heat or with metal utensils.

Sol-gel silica-based material is a coating often used on nonstick cookware as an alternative to traditional PTFE coatings. Although this coating is touted as eco-friendly and nontoxic, recent studies have raised concerns about the potential release of nanoparticles from the layer when heated to high temperatures. These nanoparticles could be harmful if ingested. The Green Pan lawsuit is one of the most notable examples of concerns around nonstick cookware coatings. Green Pan marketed their cookware as free of PFOA, PFAS, lead, and cadmium and made with a "thermalon" coating that used a sol-gel silica-based material. However, independent testing found that the cookware released high levels of toxic chemicals, including PFAS and PFOA, when heated to high temperatures. This led to a class-action lawsuit against Green Pan, which was settled for nine million dollars.

Brand recommendations: Our Place 💡, Caraway

Carbon Steel

Carbon steel cookware is a fantastic, nontoxic, sustainable option for your kitchen. It is made of iron and carbon and contains no harmful chemicals. Carbon steel is also recyclable, and its production process uses less energy than many other metals. Carbon steel heats up quickly and evenly in cooking, making it an excellent choice for searing and sautéing. It is also lightweight and durable, making it easy to handle and long-lasting. However, carbon steel does require some maintenance, as it needs to be seasoned before use and requires ongoing care to prevent rusting. It is also not nonstick, meaning you must use oil or butter to prevent sticking. Finally, while carbon steel is more affordable than other cookware options, it is still more expensive than traditional nonstick cookware.

Brand recommendations: Lodge 💡, Craft Work, De Buyer

Enameled Cookware

Enameled cookware is popular for many home cooks due to its versatility and ease of use. It's made by coating a cast iron base, stainless steel, or aluminum with an enamel layer, a type of glass fused to the metal. This makes the cookware nonreactive and easy to clean, as it resists sticking and staining. While enameled cookware can be a good option for those seeking a nontoxic and sustainable alternative, it's important to know that some enamel coatings can contain trace amounts of heavy metals, such as lead and cadmium. These metals can leach into food when the enamel is chipped or cracked. Choosing enameled cookware from reputable brands that use high-quality enamel coatings free from heavy metals is key to mitigating this risk. Avoiding cooking acidic foods, such as tomato sauce, in enameled cookware is recommended, as the acid can cause the enamel to break down and potentially release heavy metals.

Brand recommendations: Le Creuset, Great Jones 💡

Stainless Steel

Food-grade stainless steel is typically composed of 18 percent chromium and 8 to 10 percent nickel, which is why it is often referred to as 18/8 or 18/10 stainless steel. This stainless steel is considered safe for cooking and food contact as it does not leach harmful chemicals or metals into food. In terms of eco-friendliness, stainless steel cookware is not perfect, but it is considered more eco-friendly than nonstick cookware because it has no coating that can wear off or release harmful chemicals when heated. However, stainless steel cookware is still a metal and requires energy-intensive mining and processing. Some lower-quality stainless steel cookware can also contain trace amounts of heavy metals such as lead and cadmium.

Brand recommendations: Cuisinart, All-Clad, Goldilocks

Copper

Copper cookware is popular for its excellent heat conductivity and beautiful aesthetic. While copper is a natural material and can be recycled, producing copper cookware can be resource-intensive and may contribute to environmental degradation. In addition, mining and refining copper can generate pollution and waste. One potential concern with copper cookware is that it can leach copper into food, particularly when cooking acidic foods or when the cookware is scratched or damaged. Copper is an essential nutrient, but too much copper can be harmful, especially for individuals with certain health conditions. To mitigate this risk, copper cookware is often lined with another metal such as stainless steel or tin. Some copper cookware can contain trace amounts of heavy metals such as lead and cadmium, which can also leach into food.

Brand recommendations: Amoretti Brothers, Ruffoni

Glass

Glass is an excellent option. It's non-toxic and nonreactive, meaning it does not release harmful chemicals when heated or exposed to acidic foods. It is also highly durable, long-lasting, and easy to clean. Glass cookware is versatile and can be used for baking, cooking, and serving. However, there are a few drawbacks to using glass cookware: It could be a better conductor of heat, and it may take longer to heat up and cook food than other materials. It can break or crack easily, especially when exposed to sudden temperature changes. It may not be suitable for all cooktops, such as induction stovetops, and may scratch or damage certain surfaces.

Brand recommendations: Anchor Hocking, Pyrex, Thrift Shop 💡

Aluminum

Aluminum bakeware is a common and affordable option due to its excellent heat conduction. However, it is vital to be aware of some concerns about its safety, particularly with acidic foods that can cause the metal to leach into the food. While aluminum leaching is not harmful in small amounts, it can be a concern for those who use aluminum cookware frequently or have health conditions that make them more sensitive to aluminum exposure. In addition, aluminum is a neurotoxin, which is another reason to be cautious. To reduce the risk of aluminum leaching, it is recommended to avoid cooking acidic foods in aluminum cookware and to consider alternatives such as stainless steel or ceramic. Caraway offers bakeware made with aluminized steel and then wrapped in a PFAS-free nonstick coating. Many brands also provide baking sheets made with aluminum clad with stainless steel. It's crucial to ensure that whatever coating on your aluminum baking sheets is 100 percent PFAS-free!

Silicone

The safety of silicone cookware depends on the quality of the cookware you use and how you use it. In 1979, the US Food and Drug Administration (FDA) determined that silicone cookware was safe for cooking. Since then, there has been no conclusive research from the FDA or Health Canada to suggest otherwise as long as you are not subjecting it to temperatures higher than 428°F.

That said, low-quality silicone cookware can still contain additives and binders. So when buying silicone cookware, you want to look for high-quality, 100 percent pure food-grade silicone.

There are two main ways to judge if your silicone cookware is good quality. The first is the smell. High-quality silicone shouldn't have a strong odor. If your silicone item has a plastic smell out of the package, it could be better quality. The second is touch. Low-quality silicone items tend to be rougher and develop white streaks when bent and distorted. When silicone is of high quality or "food grade," it should not contain any ingredients that would leach into our food like Teflon/PTFE and other materials. Although the FDA and Health Canada have both advised refraining from using silicone for cooking past these temperatures... better to be safe than sorry! Also, good quality silicone doesn't contain BPA (Bisphenol A) or other toxic chemicals like plastic cookware products do. This is huge since exposure to BPA can influence human cell repair, fetal growth, energy levels, reproduction, and more.

Brand recommendations: William Sonoma, Stasher, Food52

Buy Secondhand

Shopping for secondhand pots and pans is an eco-friendly and economical option. You can find high-quality cookware at thrift stores, garage sales, and online marketplaces like Craigslist, Facebook Marketplace, or eBay. When shopping for used cookware, scrutinize it for any damage or wear and tear. Check for signs of rust, pitting, and scratches, which can affect the quality of the cookware and potentially release harmful chemicals into your food. You can also look for well-known brands for their durability and longevity.

As we simmer down our journey of cooking up a green kitchen, remember that the key ingredients for a more eco-friendly space are at your fingertips. These tips I've gathered and put into practice over my thirty years in this business will help you whip up a sustainable culinary haven. Be bold about embracing reusable containers, composting, and energy-saving appliances. Your kitchen will be a place where delicious meals are created and a shining example of your commitment to protecting our planet. So, put on your apron, roll up your sleeves, and let's continue to make our kitchens the heart of a greener home.

CHAPTER 3

Squeaky Green: Navigating the World of Natural Cleaners

In my childhood during the vibrant '80s, I vividly recall watching *Growing Pains*, indulging in TV dinners, and experiencing an era dominated by processed foods and products designed to make our lives easier. The fragrances of bleach and ammonia wafted through the house, becoming emblematic of cleanliness in my mind. But little did I know these products, crafted with a complex chemical concoction of ingredients, could baffle even the most seasoned scientists. My journey toward eco-conscious cleaning started in the cleaning aisle after stumbling upon an eye-opening article about unregulated chemicals in popular cleaning products.

This revelation transformed how I approached cleaning, sparking a green revolution in my home.

Unlike food, cleaning products often have no ingredient list on the label. In addition, companies are not required to disclose their ingredients, leaving consumers in the dark about what they're buying. What's always been a shock for me is that the EPA only requires companies to list active disinfectants of "known concern or potentially harmful." Governments in Canada and the USA do not test cleaning products and do not mandate that companies do either. For the average person, navigating this lack of transparency can be incredibly frustrating.

Adding to the confusion is the rampant greenwashing in the industry. Terms such as "organic," "natural," "nontoxic," and "green" are frequently tossed around, making it challenging to discern genuinely eco-friendly products from those that are merely capitalizing on the trend. This marketing tactic misleads consumers and undermines the efforts of those genuinely committed to a greener lifestyle.

COMMON GREENWASHING TACTICS

1. Ambiguous language and buzzwords:
   * A product labeled "natural" without specifying which ingredients are natural.
   * A laundry detergent marketed as "eco-friendly" without information on specific environmental benefits.

2. Misleading packaging:
   * A cleaning product with images of forests and waterfalls but containing harsh chemicals.
   * A dish soap bottle designed to look like recycled materials without explicit confirmation.

3. Exaggerated claims:
   * A product claiming to be "100 percent biodegradable" but taking a long time to degrade.
   * A disinfectant spray is promoted as "carbon neutral" but only refers to the manufacturing process.

4. Hidden ingredients:
   * A cleaning product listing "fragrance" without disclosing specific chemicals.
   * A floor cleaner uses "sodium laureth sulfate" without explaining its potential harm.

The term "green" cleaning can be vague, encompassing many interpretations. For some, "green" implies prioritizing safe and gentle products for humans, while for others, it means emphasizing sustainable sourcing or minimizing environmental impact. This diversity in understanding highlights the need for more precise guidelines and better communication to help consumers make informed choices.

COMMON PRODUCT LABELS INCLUDE

- **Cruelty-free:** Implies products were not tested on animals but might contain animal-derived ingredients. (More on this in Chapter 6.)
- **Vegan:** Indicates products do not contain animal-derived ingredients and are not tested on animals, adhering to ethical and cruelty-free practices. (More on this in Chapter 6.)
- **Eco-friendly:** Refers to products that are not harmful to the environment, indicating that everything from production to packaging is beneficial and safe for the planet, thereby minimizing environmental impact.
- **Green:** Suggests products are safer for both human health and the environment.
- **Sustainable:** A product maintains environmental, economic, and social benefits throughout its life cycle.
- **Natural:** Implies it's made from ingredients sourced from nature.
- **Nontoxic:** This signifies that the components used in products or substances are toxin-free and don't cause adverse health effects.
- **Palm oil–free:** Indicates products do not contain palm oil, an ingredient linked to deforestation and habitat loss. (More on this in Chapter 6.)

- **Sustainably sourced:** Implies that the materials and ingredients used in the product are responsibly and ethically obtained, ensuring minimal negative impact on the environment and communities. (More information in the next chapter.)
- **Biodegradable:** The problem with cleaning products labeled as biodegradable is that the term can be misleading or deceptive. Biodegradable claims are only sometimes regulated or enforced, leading to a wide variation in the actual environmental impact of these products.

Some cleaning products labeled as biodegradable might still contain harmful chemicals that can pollute water sources, harm aquatic life, or pose health risks to humans. Moreover, the rate at which these products break down can vary significantly, and some may require specific conditions, such as high temperatures or sunlight, to decompose effectively. In some cases, biodegradable cleaning products may not break down entirely in a natural environment, leaving behind microplastics or toxic substances that can still cause harm to ecosystems and human health.

Since these labels can be vague and sometimes tricky, looking for products with third-party certifications is advisable.

- **Green Seal:** a nonprofit organization that certifies products and services based on environmental standards covering the product's entire life cycle, from raw material extraction to disposal. Standards consider performance, health, and sustainability. To attain certification, products undergo rigorous evaluation, including on-site audits. Once certified, regular monitoring ensures continued compliance.
- **ECOLOGO:** A voluntary certification indicates that a product meets comprehensive, lifecycle-based environmental standards. The certification covers a range of products, including cleaning supplies, paper goods, personal care items, and toys, focusing on environmental toxins. Products undergo thorough scientific

testing and auditing to ensure compliance with these rigorous third-party standards.

- **USDA BioPreferred:** Initiated by the USDA, it promotes the use and market growth of products made from renewable resources, such as plants, to decrease reliance on petroleum. This program has two main components: mandatory purchasing rules for federal agencies and a voluntary product certification and labeling system. The program's label ensures consumers that the product has a verified amount of renewable content.

- **Cradle to Cradle:** A sustainability label for products designed for the circular economy. Founded on principles from the book Cradle to Cradle: Remaking the Ways We Make Things, it evaluates products in five areas: material health, material reuse, renewable energy and carbon management, water stewardship, and social fairness. Products receive ratings from Basic to Platinum in each category, with details publicly accessible on the organization's website.

- **EWG Verified™:** The Environmental Working Group (EWG) is a nonprofit focused on chemical safety in consumer products. Their EWG Verified mark on over 1,700 products stems from their Skin Deep database, which details ingredients in over 70,000 personal care items. Products with the EWG Verified label adhere to the organization's stringent health standards, avoiding any ingredients deemed "unacceptable" due to health or environmental concerns. This label also mandates full ingredient transparency.

- **EcoCert:** An organic certification and inspection body founded in France in 1991. Active in over eighty countries, it is one of the world's largest and most recognized organic certification organizations, mainly known for its "natural and organic cosmetic label." For a product to be certified, a minimum of 95 percent of its plant-based ingredients must be natural, and at least 10 percent of all ingredients by weight should originate from organic farming. The standard also mandates environmentally friendly manufacturing processes and prohibits using GMOs, parabens, phenoxyethanol, nanoparticles, silicon, PEG, synthetic perfumes and dyes, and certain animal-derived ingredients.

- **Scientific Certification Systems:** Ensures that products labeled as biodegradable meet the stringent standards set by the Organization for Economic Co-operation and Development (OECD) to guarantee their environmental safety and effectiveness in breaking down.

- **MADE SAFE:** This certification ensures that products, from apparel and bedding to personal care and childcare items, are devoid of over 6,500 harmful chemicals. To earn this certification, companies provide a comprehensive ingredient list for evaluation. Initially, the list is cross-referenced against banned chemicals. Subsequent assessments involve a toxicant database screening to authenticate certain chemicals. Before a final report is issued, further evaluations encompass aspects like bioaccumulation, environmental impact, and potential human health effects.

- **Safer Choice Label:** This label is an Environmental Protection Agency (EPA) initiative to identify cleaning products that meet specific safety standards. These standards encompass several criteria, including packaging, pH levels, performance, use of volatile organic compounds, and ingredient safety. However, the label has sparked some controversy due to its allowances for fragrance ingredients. Many of these ingredients pose potential health risks but are included in products primarily for their scent rather than as a functional necessity. As a result, the "Safer Choice" label could be fooling consumers who interpret it as an assurance of comprehensive safety.

The EPA introduced a separate "fragrance-free" label to address this concern. This label certifies that a product does not contain fragrance chemicals and only includes ingredients from the Safer Chemical Ingredients List, ensuring a higher level of consumer safety. Therefore, seeking out products with the "fragrance-free" label is recommended for those concerned about potentially harmful chemicals.

Beyond the surface of labels and claims lies a deeper concern—the many issues associated with traditional cleaning practices. From harmful toxins silently infiltrating our living spaces to the significant environmental impact of conventional cleaning routines.

BEYOND THE BOTTLE: CLEANING'S HIDDEN IMPACT

Chemicals

Maintaining a clean home doesn't mean it should have a particular scent. The greatest challenge for those transitioning to natural cleaning methods is the absence of fragrances. Marketing messages from companies selling their products have led us to believe that a clean home or freshly laundered clothes should have a distinct smell. However, the reality is that cleanliness has a variety of aromas.

Adding fragrances to cleaning products is a clever marketing tactic to evoke positive emotions and persuade customers that their homes are clean based solely on the scent, resulting in a loyal consumer base. But the truth behind these fragrances is far from appealing.

Fragrances are created using a combination of synthetic chemicals designed to produce a unique scent. While chemicals are not inherently harmful, ensuring their safety is crucial, which requires disclosing all ingredients. Unfortunately, full disclosure is often not the case regarding fragrances. The term "fragrance" is considered a trade secret, so companies aren't obligated to reveal the chemicals or formulas used in their products. Fragrance is not an ingredient. It's just a vague term.

Some certifications have been exposed for greenwashing or fraudulent practices. Conversely, there are businesses genuinely committed to environmental sustainability but need more resources to invest in expensive, well-known certification programs.

Ultimately, more than certifications are required to ensure ethical and eco-friendly practices. That's why I encourage you to research and hold brands accountable.

Understanding the types of chemical ingredients found in cleaning products is crucial for making informed decisions about the products we use in our homes. Speaking of chemicals, let's clear up some confusion.

To fully understand the concept of toxicity, it's important to differentiate between toxins and toxicants. Toxins are naturally occurring harmful substances, such as those in poisonous mushrooms or snake venom. In contrast, toxicants are artificial, man-made substances that can harm health, often resulting from human activities like industrial waste production and pesticide use. Many chemicals in our homes tend to be toxicants rather than toxins.

Again, not all chemicals are inherently harmful; they form the foundation of all matter. Therefore, using phrases like "chemical-free" when referring to nontoxic products is deceptive. Our living spaces can contain various toxicants that pose health risks. For example, we consume food and water tainted with hormone-disrupting pesticides, BPA, and dioxins, which accumulate in our bodies and cause health issues.

We inhale harmful chemicals associated with asthma and allergies, such as artificial scents in personal care, cleaning items, and phthalates from dust and construction materials.

Chemicals in laundry and skin care products can also affect our skin, such as sensitizing ingredients like methylisothiazolinone and hormone disruptors like phthalates and parabens. The body's toxic load encompasses the total amount of toxins accumulated over time from various sources, including air, water, food, and everyday products, which consist of heavy metals, pesticides, chemicals, and other harmful substances. These substances can disrupt normal bodily functions, resulting in various health problems. Studies indicate that even low-level exposure to toxins can contribute to chronic diseases like cancer, neurological disorders, reproductive issues, and autoimmune diseases. Specific groups may be more susceptible to toxicant effects, including children, pregnant women, and individuals with preexisting health conditions.

Indoor Air Pollution

You've just finished tidying up your home, wiping down surfaces and floors with your go-to cleaning products. The room smells fresh, and everything sparkles like new. But what if I told you that the same products you trust to keep your home clean secretly contribute to an invisible enemy—air pollution?

That's right! Many conventional cleaning products contain a cocktail of chemicals that, when used, release harmful pollutants into the air we breathe. These volatile organic compounds, or VOCs, are like silent ninjas, infiltrating our homes and impacting our health without us even realizing it.

As you spritz and scrub away with your trusty cleaning products, VOCs evaporate into the air, forming a toxic cloud that can irritate your eyes, nose, and throat, trigger allergies, and even cause serious long-term health issues like respiratory problems and cancer.

But wait, there's more! These sneaky VOCs don't just wreak havoc indoors; they also contribute to outdoor air pollution when they escape our homes through windows and ventilation systems. Once outside, they mix with other pollutants and create smog, further exacerbating the air quality crisis.

The list below highlights fourteen ingredients to watch out for.

CLEAN WITH DELIGHT, FOURTEEN TOXINS TO AVOID ON SIGHT

1. 2-Butoxyethanol (2-BE): Colorless solvent used in cleaning products, potentially harmful to health and the environment.
2. Coal tar dyes: Synthetic dyes used for coloring can contain heavy metals and pose a cancer risk.

3. MEA (monoethanolamine), DEA (diethanolamine), TEA (triethanolamine): Ethanolamines in cleansing agents may be contaminated and linked to health issues.
4. Nonylphenol ethoxylates (NPEs): Surfactants in detergents associated with hormone disruption and harm to aquatic life.
5. Fragrance/parfum: Commonly used scent chemicals may contain phthalates and contribute to indoor air pollution.
6. Phosphates: Used as water softeners in cleaners, can lead to algal blooms and harm aquatic ecosystems.
7. Sodium lauryl sulfate (SLS) and sodium laureth sulfate (SLES): Surfactants found in various cleaning products; potential eye and skin irritants with environmental concerns.
8. Sodium hydroxide (lye and caustic soda): Strong alkaline compound used in cleaners can cause burns and environmental damage.
9. Sodium dichloroisocyanurate dihydrate: Chlorine bleaching agent in disinfectants; highly toxic to humans and aquatic life.
10. Triclosan (TSC): Antimicrobial agent in cleaners; suspected endocrine disruptor with potential antibiotic resistance concerns.
11. Ammonia: Pungent gas used in various cleaners can irritate the respiratory system and harm aquatic life.
12. Silica powder: Abrasive ingredient inhalation can be dangerous and classified as a human carcinogen.
13. Trisodium nitrilotriacetate: Water softening agent, classified as a possible human carcinogen and harmful to aquatic life.
14. Quaternary ammonium compounds (Quats): Used in disinfectants and cleaners, they may cause skin irritation and contribute to antibiotic resistance.

Toxic Water

Water pollution caused by cleaning products is often overlooked. Our everyday use of household cleaning products contributes to water pollution.

Many harmful ingredients go down our drains and into the sewer system, wastewater treatment facilities, and ultimately our rivers, lakes, and oceans.

Wastewater treatment systems are designed to break down chemicals before they enter the environment. However, they struggle to handle all the harmful chemicals in cleaning products. Consequently, these chemicals often end up in our freshwater and saltwater ecosystems, severely threatening animals, plants, and our drinking water and health.

A 2002 United States Geological Survey study found detergent traces in 69 percent of streams across the USA. Moreover, the Environmental Working Group's (EWG) Tap Water Database discovered more than 250 chemicals in America's drinking water. Astonishingly, over 160 contaminants have no governmental limits set.

Never toss cleaning products in the trash. Many are considered household hazardous waste and must be disposed of responsibly. Look for drop-off depot or community environment days in your area.

Plastic Crisis

When discussing plastic waste, particularly concerning plastic bottles, water and soda bottles often come to mind as the main offenders. However, they are not the only contributors. Non-reusable cleaning and laundry product bottles also significantly contribute to landfills worldwide. Between 1950 and 2017, 7 billion tons of the 9.2 billion tons of plastic produced became plastic waste. That's enough plastic to fill 3.68 million Olympic-sized swimming pools.

Sadly, a large portion of this plastic waste finds its way into our oceans, lakes, and rivers, posing a severe threat to the health and livelihoods of millions of people globally. We are facing a plastic crisis, and proper recycling is crucial.

The recycling triangle symbol with a number in the middle on the bottom of plastic items like laundry detergent bottles is known as the Resin Identification Code (RIC). It doesn't directly indicate how to recycle the plastic but identifies the type of plastic resin used. This information helps recycling facilities sort and process plastics appropriately for recycling or disposal. The most readily recyclable plastic numbers are #1 and #2, typically accepted in curbside pickup programs. In addition, consider using specialized recycling services for other plastic types to ensure proper disposal. The production of plastic cleaning bottles derived from fossil fuels and their transportation contributes to greenhouse gas emissions, exacerbating global warming. I bet you are surprised to learn about the connection between climate change and the act of cleaning your homes.

CLEANING UP YOUR CLEANING ROUTINE

Got enough dirt up to this point? Okay, let's move on. If you asked me twenty years ago if you should toss out all your conventional cleaners, I'd probably say yes. But times have changed, my friends—no need to do that. Just finish them off and start replacing them with "better" ones. By "better," I mean what's better for you. My advice, which I share on The Eco Hub, is to follow any or all of these seven pillars of sustainability:

1. **Better Ingredients:**

    * Prioritize products with natural ingredients. Refer to chart.

    * Steer clear of GMOs, preservatives, parabens, and toxicants.

    * Refrain from buying petroleum-based products.

    * Examine sourcing practices to avoid products with chemical herbicides, pesticides, and artificial fertilizers.

    * Even if products are labeled organic or natural, they might still be tested on animals, contain animal by-products, or include palm oil.

2. **Vegan:**

    * Choose products that exclude animal ingredients. If present, ensure they are ethically sourced, like beeswax.

    * Understand that "vegan" doesn't always mean "cruelty-free." Some vegan beauty products may still be tested on animals.

* Be aware that vegan products might contain palm oil, a key contributor to deforestation and habitat destruction.
* Remember that nonorganic or non-natural vegan products can harm the environment and animal habitats.

3. Cruelty-free:

* Seek out products with no animal testing history.
* A cruelty-free label does not automatically guarantee the absence of animal ingredients or environmentally harmful components.
* Always cross-check other certifications and the cruelty-free tag to ensure a holistic approach to ethical consumption.

4. Palm oil–free:

* Opt for products that are devoid of palm oil and its derivatives.
* If a product does contain palm oil, ensure it's sourced sustainably.
* Recognize that sustainable palm oil can sometimes be the better option as it promotes broader industry change and prevents challenges tied to alternatives that might also lead to deforestation.
* However, even palm oil–free products might contain other nonorganic ingredients or be tested on animals.

5. Sustainable Sourcing/Ethical Manufacturing:

* Emphasize ethical sourcing when selecting products.
* A company's dedication to ethical sourcing indicates a holistic view, valuing the ingredients and those who produce them.
* Review Fair Trade labels and thorough company research to gauge a brand's sustainability efforts.

6. Eco-friendly Packaging:

* Prioritize reducing waste for the planet's well-being.
* Opt for brands that promote packaging reuse or offer compostable solutions.
* Brands using post-consumer recycled materials and easily recyclable packaging (like glass or aluminum) should be preferred.
* At the baseline, ensure the packaging is recyclable and, if possible, made from recycled materials.

7. Septic-safe:

* Choose cleaning agents that decompose swiftly to avert damage to septic systems. This supports a balanced bacterial environment in the septic tank, facilitating efficient waste decomposition and protection against blockages or pricy repairs.

> I'm sharing my favorite cleaning products at the end of this section. Remember to refer to the above criteria. Note the criteria are based on the company as a whole, not the individual product.

You don't have to follow all of the above; some might be easier to attain than others. Once you have established which criteria matter to you, it's time to come to terms with the fact that we do not need a cleaner for everything from the toilet to the kitchen sink. Well, maybe in those two cases, we do! LOL. But you get my drift. The art of cleaning your home lies in its simplicity.

In marketing, social engineering manipulates people's emotions and behaviors to convince them they need specific products. I've coined "con-venience" to highlight this manipulative aspect, suggesting consumers are "conned" into believing these products are necessary. Often, people prioritize quick solutions over more sustainable alternatives due to this perceived convenience. For example, Windex and Pine-Sol have successfully made strong brand associations as window and floor cleaners, respectively, by employing such marketing techniques.

This is especially true in the case of disinfectants. We are overusing them, and the results are not good. While antibiotics and antibacterial cleaning agents are vital when truly needed, our overuse creates "superbugs" immune to antibiotics and kills off our "good" gut bacteria. The same happens when you wage germ warfare on every surface of your home.

According to the CDC, "The relatively recent increase of surface antibacterial agents (found in soaps, dishwashing detergents, and all-purpose sprays) or biocides into healthy households may contribute to the resistance problem." Avoid products labeled "disinfectant," "sanitizer," or "antibacterial." Use soap and water; if you need to disinfect, use hydrogen peroxide (just add a spray pump, and you're done) or 70 percent alcohol. There is undoubtedly a place for these products but not for everyday healthy households.

Bleach has historically been a trusty weapon in the fight against germs, renowned for eradicating all microorganisms it comes into contact with. Hence, it's unsurprising that its demand soared amid a global health crisis. However, despite the indisputable germ-killing prowess of bleach, it has raised serious health and environmental concerns, given the current high sales volume.

Typical household bleach is a chemical cocktail, with sodium hypochlorite being its primary and most contentious ingredient. First used in the eighteenth century as a disinfectant, stain remover, and fabric lightener (earning its name), bleach is still widely used for these purposes today. However, our growing awareness of its potential harm to human health and the environment warrants caution.

Bleach, a highly reactive corrosive substance, emits lingering, irritating fumes. Mitigating its adverse effects is possible with appropriate measures, such as dilution with water, avoiding combination with other chemicals, and use in well-ventilated areas. Nonetheless, more than these measures may be required for specific individuals, such as asthmatics, who may find its impact significantly more problematic.

Equally concerning are the environmental consequences of bleach use. Being an organochlorine, bleach is foreign to nature. Its power as a disinfectant lies in the inherent instability of chlorine, which readily bonds with other chemicals to eliminate microorganisms. However, this same instability poses a threat when it interacts with various other chemicals in the environment.

Bleach inevitably infiltrates our environment, both through waterways and the atmosphere. Bleach can generate carcinogenic dioxins in rivers, posing a severe risk to aquatic wildlife and humans. Atmospheric bleach is linked to ozone depletion, which brings about lasting environmental damage.

While household use contributes some amount of bleach to the environment, such as when flushing a toilet cleaned with bleach, the primary source of environmental harm stems from industrial manufacturing and large-scale use. Therefore, even our individual, seemingly insignificant usage supports a destructive industry.

For years, mainstream advertising has fueled our fear of bacteria, leading to the belief that only a hefty dose of bleach can keep our homes safe. This fear has driven the toilet "care" industry, which heavily relies on bleach, to become a multi-million-dollar enterprise. However, bleach manufacturers have defended their products, arguing that the low concentration of chlorine bleach traces in wastewater does not form toxic by-products.

However, a pressing ethical issue surrounds the manufacturing process. Bleach is a member of the organochlorine family, a group of rarely naturally occurring compounds that can take centuries to decompose. Traces of these compounds have been found in environments such as America's Great Lakes and even in human breast milk—Greenpeace advocates for a complete cessation of organochlorine production. Thus, purchasing bleach inadvertently supports this harmful industry.

We must ask ourselves if the job that bleach does is essential. As we navigate the "Age of Bacteria," we should consider the perspective of immunologist Gerald N. Callahan, who suggests that completely eradicating microorganisms is not beneficial for either side. Instead, we should strive to coexist in harmony. As such, we must carefully consider our choices and their implications.

Cleaning up your act and switching to better cleaning products is easier than you think. My advice is simple: make better choices as you run out of a particular product. Try different brands; make it a permanent switch if they work for you. Or start with the products you're least picky about. I also recommend focusing on multipurpose products to simplify your cleaning routine.

You can also follow my four-step formula when shopping for better cleaners:

1. Does the item have a complete ingredient list? No list? Put it back! If they don't want to share the list, move on.
2. Avoid products that have words like flammable, explosive, corrosive, hazardous, or poisonous. Looking at you, bleach!
3. If the word fragrance or parfum is on the list, put it back on the shelf.
4. If the product passes the above, take it one step further by digging deeper into the ingredient list.

Navigating the path to healthier brands may be challenging due to the ever-evolving market. Therefore, it's vital to grasp the essentials of label-reading to decide whether a product aligns with your needs rather than relying solely on specific product suggestions.

Many conventional stores now offer eco-friendly sections. However, remember that these sections may not always guarantee stringent ingredient standards, so it's essential to continue reading labels.

Apps for reading labels have been a huge help in my journey toward choosing healthier products. However, it's good to keep a few things in mind when using them. First, the science they use is often broad and might only sometimes consider how we use products. Second, a product's overall rating only sometimes tells the whole story about each ingredient. So, while these apps are super handy, remember they are tools to help you make more informed choices, not the final word.

Also, you've probably read somewhere that avoiding an ingredient is best if you can't pronounce it. This is not always the case. Some of the simplest ingredients have the most compliant names. For example, Melaleuca alternifolias is Tea Tree Oil.

In the world of eco-friendly cleaning, innovations such as refillable products, solid bars, and concentrates have significantly shifted how we approach household chores. One of the key benefits of refillable cleaning products is their capacity to reduce packaging waste dramatically. Instead of discarding a plastic container every time a product runs out, we can refill it, cutting down on single-use plastics and the carbon footprint associated with their production and disposal.

Solid bars, on the other hand, offer an innovative alternative to liquid soaps and detergents. They last longer and eliminate the need for plastic bottles, again reducing waste. Moreover, given their smaller size and weight, they require less water in their production process and less energy to transport because they're more concentrated than their liquid counterparts.

Concentrates take this a step further. These powerful cleaning agents are sold in smaller quantities because they're designed to be diluted at home with water. This means fewer resources are used in packaging and transportation, and we, as consumers, get more bang for our buck. A single bottle of concentrate can often replace multiple bottles of a traditional cleaning product.

The common thread tying these innovations together is their ability to reduce waste, conserve resources, and minimize environmental impact without sacrificing the quality of cleaning.

Product Recommendations

For full reviews on the following, go to The Eco Hub online.

ALL-PURPOSE CLEANERS

Meliora All-Purpose Cleaner

Attitude Home Essentials All-Purpose Cleaner
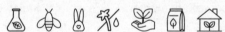

Elva's All Naturals 1 CLEANER All in One
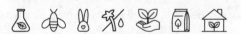

Blueland Cleaning Sprays
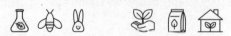

DISH SOAP

No Tox Life Dish Block

Dropps Dishwasher Detergent

Dr. Bronner's Pure Castile Soap 💡

BATHROOM CLEANERS

Branch Basics Concentrate

AspenClean
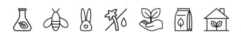

The Unscented Co.

TUB AND TILE CLEANER

Grove Co/ Tub and Tile Concentrate

AspenClean Green Powder Cleaner SuperScrub

TOILET BOWL CLEANER

Etee's Probiotic Toilet Bowl Cleaner

The Refill Shops Toilet Tablets

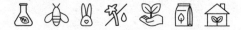

Building Your Green Cleaning Caddy

CLEANING PRODUCTS

I'm about to let you in on this secret world of homemade cleanliness that's fun and incredibly pocket-friendly. I understand it might seem daunting, but trust me, it's as simple as pie, and you'll be thanking yourself later. Even if you're not quite ready to whip up your cleaning solutions, there's no need to worry. Let's dive into the must-have items you should stock in your kitchen that will transform how you tackle tidying up.

All-purpose cleaner: A good one is a versatile workhorse in your cleaning arsenal. It's designed to tackle many cleaning tasks, saving you time and money by reducing the need for multiple specialized products.

An effective all-purpose cleaner can clean various surfaces, including countertops, kitchen appliances, bathroom fixtures, and tiled floors. It should be able to cut through grease, remove grime, and eliminate common household stains.

In addition to cleaning, a high-quality all-purpose cleaner often has deodorizing properties, leaving your spaces smelling fresh.

Glass/mirror cleaner: to be used on any surface that you want to be streak-free.

Castile soap or Sal Suds: Castile soap is a versatile cleaner made from plant oils. It's biodegradable, safe for all skin types, and can clean almost anything from your body to your dishes and laundry. Sal Suds is a more heavy-duty cleaner. It's perfect for stubborn grime and stains but is still biodegradable and plant-based. It is my go-to for tougher cleaning tasks. Both are excellent foundations for your DIY cleaning solutions.

Vinegar: A popular choice for cleaning various household surfaces such as sinks and countertops, and can effectively eliminate odors and remove dirt.

> Understanding the difference between cleaners and disinfectants is key. Cleaners physically remove dirt, debris, and some germs from surfaces, while disinfectants destroy harmful germs upon contact. White distilled vinegar, with its 5 percent acetic acid content, excels as a cleaner by dissolving dirt and grime.
>
> However, vinegar's abilities as a disinfectant are limited. It can kill or reduce specific pathogens, like E. coli, Salmonella, and Listeria monocytogenes, known for causing common foodborne diseases. Some studies suggest a 10 percent malt vinegar solution can kill the influenza A virus.
>
> Despite these capabilities, vinegar can't eliminate all types of germs, including the virus that causes COVID-19.
>
> For effective disinfection, a product should be able to kill 99.9 percent of harmful germs within five to ten minutes. Products containing ingredients like ethanol, isopropyl alcohol, hydrogen peroxide, quaternary ammonium, phenolic compounds, and sodium hypochlorite (bleach) meet this standard and can effectively kill many pathogens. Always check for an "EPA Reg. No." label to ensure the product is a registered disinfectant.

💡 **Baking soda:** Also known as sodium bicarbonate, is the unsung hero lurking in the back of your pantry. This humble, budget-friendly ingredient wears many hats, and saving you money is one of its superpowers.

Baking soda is a jack of all trades. It's not just a vital ingredient for baking; it's also a powerful cleaner. It can scrub stubborn grime and stains, deodorize foul smells, unclog drains, scrub sinks and toilets, and even boost your laundry.

Essential oils are a fantastic addition to DIY cleaning solutions, adding a natural fragrance and additional cleaning power. In addition, they offer many properties: some have antibacterial and antiviral qualities, while others act as natural deodorizers. This means they can assist in not just making your home smell good but also in eliminating harmful germs.

Lemon and tea tree oils are trendy for their fresh, clean scents and strong antibacterial properties. With its calming fragrance, lavender is also a great choice and boasts antimicrobial benefits. Peppermint oil offers a refreshing aroma and can deter pests like ants and spiders. Eucalyptus oil is another excellent option, known for its natural disinfectant qualities and refreshing scent.

Incorporating essential oils into your cleaning routine allows you to customize the scent of your homemade cleaning products while enhancing their effectiveness.

Essential oils are highly concentrated plant extracts containing potent chemical compounds. They've been used for various purposes, including aromatherapy, personal hygiene, and health concerns. While they offer numerous benefits, using them responsibly is crucial due to their potency. Also, some essential oils can be harmful to pets and kids, so it's necessary to research which ones are safe to use around them. Finally, when purchasing, pay attention to the botanical name and origin of the oil for quality assurance.

Hydrogen peroxide is an excellent green cleaner due to its versatility and effectiveness. It's a powerful oxidizer, making it great for killing bacteria, viruses, and mold. It's safe for the environment as it breaks down into water and oxygen, leaving no harmful residues behind. It's cost-effective, readily available, and can be used on many surfaces, making it a staple for eco-friendly cleaning. It also has bleach-like properties.

Alcohol: There are primarily two types of alcohol that are effective for DIY cleaning. The first is isopropyl alcohol, also known as rubbing alcohol. With a concentration of at least 70 percent, it is a potent disinfectant, perfect for sanitizing surfaces. It's a popular choice for cleaning electronics, mirrors, and stainless steel; it evaporates quickly, leaving no streaks behind. The second is ethyl alcohol, or ethanol, in vodka and other spirits. While not as potent a disinfectant as isopropyl alcohol, it is still useful in homemade cleaners for general cleaning and deodorizing tasks, especially when combined with other ingredients like vinegar or essential oils. Just be sure to avoid "denatured" alcohol, which contains added chemicals.

Distilled water: Unlike tap water, which can contain minerals and impurities, distilled water is free of these potential contaminants. This ensures your homemade cleaning solutions are more effective and avoid unwanted residue or spots, particularly on surfaces like stainless steel. Furthermore, distilled water reduces the likelihood of anything unwanted growing in your homemade products, enhancing their shelf life and stability. This makes it an optimal choice for a spotless, streak-free shine in your cleaning tasks. It can get expensive, so I suggest only using it if you need to store your homemade cleaners for an extended period.

GREEN CLEANING CLOTHS

Paper towels are my biggest pet peeve regarding waste. Why? Because we purchase them knowing full well that they will be discarded, creating a significant burden on the environment. Don't get me wrong; I understand how difficult it is to break away from this habit. Using paper products in the kitchen, such as towels and napkins, leaves a substantial carbon footprint. It begins with the resources required for their production, which involves cutting down ancient forests.

Additionally, a considerable amount of water and bleach is needed for manufacturing. Finally, the products must be transported across the country or even the globe. You get the point! Paper towels are inexpensive and disposable, leading us to use them rapidly. And once they're gone, we hurry to purchase more, perpetuating the cycle. So, how do you get started with switching to reusable paper towels? Finish up what you have before you even consider purchasing anything new. I will not lie. Paper towels are one of the most challenging habits to break!

Step 1

- Do a paper towel audit. Take a moment to consider how you currently use paper towels in your home.
- Are you using it to clean the sink or other surfaces like glass or mirrors?
- Do you use it when cooking to absorb grease (like bacon)?
- Are you using it to wipe up gross messes from kids and pets (hello, fur ball)?
- Do you dry your dishes with it?
- Do you use it to clean a lot on the dining table or a spill in the kitchen?

Step 2

Out of sight, out of mind. I want you to move your roll away from where it's easily accessible. Put it in the pantry or under the sink. The point is to get it out of arms' reach.

Step 3

Have lots of reusable cloths on hand all the time!

Use what you have: Cloth napkins make an excellent choice for drying hands or cleaning minor spills. In addition, they're reusable and can be laundered like any other clothing. Washcloths provide another option for mopping up spills or wiping surfaces. Compact and easy to store, they can be washed and used repeatedly. Dish towels are frequently employed for various cleaning tasks, including wiping countertops or drying kitchenware. These durable items can be washed and reused multiple times as well.

Old socks, T-shirts, towels, or pillowcases can be repurposed into cleaning cloths. Additionally, if you have cloth diapers, the reusable wipes that often accompany them can substitute paper towels. These wipes, made from soft and absorbent materials, are ideal for tackling spills and messes.

The secret to utilizing available resources as paper towel alternatives lies in creativity and evaluating the materials you possess. This approach can decrease waste and expenses while maintaining a clean and orderly home.

It's worth mentioning that the absorbency and gentleness on delicate surfaces may vary among different materials. However, repurposed t-shirts and towels typically suffice for everyday cleaning tasks, such as wiping down countertops and tables.

For even greater convenience when cleaning with repurposed fabric, designate a specific area for soiled rags and launder them in a separate load. This method helps segregate them from your regular laundry, avoiding any potential cross-contamination.

Swedish dishcloths are versatile and eco-friendly cleaning tools, also known as Swedish sponge cloths, and are indispensable for maintaining a clean home. They are highly absorbent and made from cellulose and cotton, perfect for tackling messier liquid spills and oily surfaces. Incredibly practical, they can be used for various tasks, including cleaning sinks, stoves, refrigerators, and floors and washing dishes. Exceptionally durable, these reusable cloth wipes can withstand up to a hundred washes and last between five and six months, depending on usage. In addition, they are machine washable, dishwasher safe, and equivalent to fifteen rolls of paper towels. The best part? They are 100 percent backyard compostable, making them an environmentally friendly choice. Swedish dishcloths come in various sizes and are safe to use on all surfaces, so you don't have to worry about damaging your furniture, countertops, or other household items. In addition, their slim and narrow shape allows for cleaning in hard-to-reach nooks and crannies, such as between appliances, furniture, and other tight spaces where dirt and grime can accumulate.

To ensure optimal longevity, air dry them properly to prevent unpleasant smells. You can store them in a drawer or hang them over a tap, allowing them to dry with a curve. Though they may have a higher price tag than other cleaning cloths, their value per use justifies the cost. With at least six in your collection, you'll appreciate the convenience and effectiveness of these reusable Swedish dishcloths in your daily cleaning routine. By replacing paper towels with a ten-pack of Swedish dishcloths, you could pocket an annual savings of $135.

Huck towels were initially designed as surgical towels. They are 100 percent cotton, highly absorbent, and have extremely low lint, making them perfect for cleaning windows and glass. Their durability is impressive, as they maintain their absorbency and softness even after countless washes. Huck towels are versatile enough for various cleaning tasks, such as drying raw meat and tidying up after meal prep. Just rinse them with hot water and toss them in the wash afterward. Having at least twenty-four towels on hand is recommended for optimal cleaning coverage. You can purchase either new or reclaimed Huck towels, but I highly suggest going for the reclaimed ones. Buying them in bulk can save you money, and once you've experienced their cleaning prowess, you'll never want to go back!

Reusable cloth wipes are made from soft, durable materials like cotton flannel or bamboo. They can be used for various purposes, including wiping down surfaces, cleaning up spills, or even during camping trips, picnics, and kids' lunch boxes. In addition, they help reduce waste and avoid harsh chemicals often found in commercial disposable wipes. The soft texture of the cotton flannel renders these cloths gentle on delicate surfaces, such as glass and electronics, which has proven invaluable in many households. Highly absorbent, they pick up dirt, dust, and other particles, making cleaning spills and messes a breeze. The durability of these cloths is impressive, as they hold up well despite repeated use and washing.

However, they may be less effective for greasy or oily messes and might need washing after each use. Also, improper washing and drying can cause shrinkage or fading over time, and bacteria buildup is safe if thoroughly cleaned.

Unpaper towel rolls are available with or without snaps or buttons and are made from cotton, bamboo, or flannel materials. They are durable, absorbent, and easy to clean. These cloth sheets can be used for various tasks like wiping surfaces, cleaning spills, or drying hands. Unpaper towels with snaps or buttons can be connected to form a roll, making them convenient to tear off individual sheets as needed and store on a standard paper towel holder. However, one potential drawback is that they require more effort and time to fasten and unfasten than traditional paper towels, which can be inconvenient when dealing with spills or messes that need quick attention. In addition, after an extended period, around five years, the buttons may become worn, making the towels less secure and less effective at staying in place. When using these unpaper towels, it is essential to be cautious about scratching sensitive surfaces like wood and glass. You may need to fold the cloth back to avoid the buttons touching any surfaces, making them suitable for tubs and tiles but requiring caution on more sensitive areas.

Unpaper towels without snaps or buttons are simpler in design, stored in a stack, or folded individually. They don't require assembling or disassembling the roll and don't pose the risk of scratching surfaces. However, they may need a designated storage space or container since they can't be stored on a standard paper towel holder. Both unpaper towels are reusable and washable. The choice between the two types depends on personal preference for storage, convenience, and aesthetics.

How I Use All of the Above

For ultimate cleaning success, I make plenty of reusable cloths readily available in key areas of my home. I take organization one step further by placing baskets in these strategic locations, with one labeled "clean" and the other marked "dirty." This system ensures my family knows exactly what to do with the used cloths. I keep a mesh bag for collecting dirty cloths in the kitchen and bathroom cupboards. When it's time for laundry, I toss the bag into the washing machine. For more heavily soiled or greasy cloths, I presoak them in baking soda before adding them to the wash. I keep the kitchen and bathroom cloths separate when washing. I use color coding, for example, green for the kitchen and yellow for the bathroom.

SPONGES

Out with the old, in with the new—yellow and green sponges may give the illusion of a squeaky-clean home, but they leave a dirty mark on Mother Earth. Made from non-biodegradable materials like plastic and petroleum-based foams, these sponges take eons to decompose, adding to the mountain of plastic pollution. Moreover, as they scrub-a-dub-dub away grime, they shed microplastics into our waterways, creating a ripple effect of harm to marine life and ecosystems.

Instead, try the following:

Non-abrasive scrubbers: These are fantastic for cookware, ceramic stovetops, sinks, tubs, tiles, and more. They are biodegradable and tough on dirt. I keep two of them in my bathroom. They are great for helping remove mildew that can develop around the bathtub.

Coconut husks: Boy, do I love these! They are made from coconut husk! I have both the brush and the scours. The brush is super durable and can scrub pots, pans, dishes, and hard-to-reach places. The hands are made from FSC-certified wood, and the bristles are made from sustainably farmed coconut husk, the outside of dried coconuts! The scrubber is ideal for cleaning pretty much anything in your home.

Both are naturally antibacterial, easily remove mold, and perfect on hard surfaces.

Other notable zero-waste cleaning tools are brushes from Burstenhaus Redecker. It comes with a long handle and it's made with natural fiber bristles, perfect for cleaning pots and pans. It works hard at scrubbing those hard-to-clean sticky food residues. It comes with a replacement head, so hang onto the handle. I've had mine for two years and have yet to replace it. I also use and love my dish scrubber. It's small and significant with block soap. It's ideal for pots and pans.

Pot Scraper: This is the best thing I have ever bought. You don't scrub, you scrape. Let the pan cool, then scrape off all the excess food. There's no need to scrub, and you will use less soap and water too.

EQUIPMENT

Cleaning equipment such as vacuums, mops, buckets, large and small brooms, and dustpans are essential to maintaining a clean and healthy living space. These tools are designed to handle different cleaning tasks effectively. For example, vacuums are excellent for removing dust and debris from carpets and upholstery, while brooms and dustpans handle dry messes on hard surfaces. Mops and buckets are indispensable for washing floors, helping to remove dirt and sanitize.

Now that you know how to select the best eco-friendly cleaning products, are you ready for an exciting twist? It's time to roll up your sleeves, tap into your inner scientist, and embark on a DIY journey. Let's venture into the thrilling world of creating your cleaning solutions right in your home!

The Art of DIY Cleaners

Maintaining hygiene and safety while preparing homemade cleaners is essential. Always use clean containers, spray bottles, and utensils during preparation. Dirty tools can introduce bacteria into your DIY products.

When creating long-lasting cleaners, choose distilled water available in stores or boil tap water for at least fifteen minutes and let it cool. Distillation or boiling helps remove contaminants, reducing the chances of bacterial growth in your cleaning solution. For short-term use, filtered tap water is suitable.

Pay attention to the condition of your DIY cleaners. Any signs of contamination, like an off smell, black mold, unusual growths, or a slimy film, indicate it's time to dispose of that batch and make a new one.

Remember, not all surfaces react the same way to a cleaning product. The type, age, and surface coating can all influence the outcome. Therefore, always perform a spot test on a hidden area to avoid unwanted results. This is especially crucial for valuable or delicate surfaces like antiques, hardwood floors, or countertops.

Lastly, label your products clearly with the name and date. It can be as simple as using tape or a permanent marker on glass spray bottles. This practice will prevent any mix-ups between similar-looking liquids.

You wouldn't want to water your plants with a vinegar spray, mistaking it for plain water.

Let's dive into the world of natural cleaning ingredients and understand the combinations to steer clear of. Interestingly, the element that usually stirs up trouble in these pairings is vinegar. It's the proverbial black sheep among natural ingredients that seems to clash with others, particularly hydrogen peroxide, castile soap, and baking soda.

Vinegar, although harmonious with many ingredients, has the potential to trigger unwanted chemical reactions. Let's understand why:

1. **Vinegar + Baking Soda:** This duo has been heavily endorsed online and in numerous cleaning books. However, combining vinegar (acidic) and baking soda (alkaline) results in a neutralization reaction, leaving a saltwater solution with minimal cleaning prowess. You might still notice a clean drain after using this mix, probably due to the reaction occurring within the drain. Use these two sequentially for cleaning tasks but avoid combining them into one solution.

2. **Vinegar + Castile Soap:** This combination may not pose a risk, but it does result in an ineffective oily mix. Vinegar and Castile soap neutralize each other, reducing the soap to its initial oil form, rendering it ineffectual. Sal Suds, a detergent differing in composition from Castile soap, can be used as a viable alternative to vinegar.

3. **Vinegar + Hydrogen Peroxide:** Merging these two powerful disinfectants produces peracetic acid. Despite low toxicity, it is corrosive and can irritate the skin and eyes, potentially triggering asthma. Instead of blending them into one solution, use hydrogen peroxide and vinegar sequentially for disinfection or a pure hydrogen peroxide or vinegar spray.

SPARKLE SPRITZ ALL-PURPOSE CLEANER

Ingredients:
1 16oz glass spray bottle
2 cups distilled or filtered water
2 tablespoons of liquid Castile soap
15 drops essential oils (optional)

How to make:
Using a funnel, pour the water into the bottle.
Add the Castile soap.
Add the essential oils.

How to use:
Shake bottle before you spray.
Use like any all-purpose cleaner.

Where to use:
Bathroom: sinks, tubs, counters, stainless steel appliances
Kitchen: counters, sinks, toilets

GLASS AND WINDOW WONDER

Ingredients:
1 16oz glass spray bottle
1 cup distilled or filtered water
3 tablespoons vinegar
¼ cup rubbing alcohol

How to make:
Using a funnel, pour vinegar and alcohol into the bottle.

How to use:
Shake the bottle before you spray.
Spray onto a cleaning cloth.
Wipe the surface.

Where to use:
Glass and mirrors

ECO-CLEAN MAGIC SCRUB

Ingredients:
1 cup baking soda
¼ cup Sal Suds
1 tablespoon hydrogen peroxide

How to make:
Combine all the ingredients in a bowl.
Mix until a paste forms.
Transfer the scrub into a dark-colored, airtight jar (leave at least 2 inches at the top to allow for expansion).

How to use:
Apply a small amount of paste to a cloth or brush.
Use the cloth to clean the surface.
Leave for a few minutes, then wipe or rinse.
Store in a cool dark place—lasts for 1 month.

Where to use:
Tubs, tiles, showers, and sinks

In this chapter, I've taken you on a journey through eco-cleaning, showing you how we can create a significant impact with small, sustainable changes in our housekeeping routines. We've swapped out harsh chemicals for natural, non-toxic ingredients and exchanged single-use items for reusable ones. Through this, we've found that we can keep our homes sparkling clean while also contributing to the *sustained* health of our planet. By making these conscious choices, we're playing an active part in a greater, *sustained* environmental effort that reaches toward our shared dream—a preserved Earth for our children and our children's children. In our exploration of eco-cleaning, we've seen firsthand that every single action in our daily lives, no matter how simple or routine, can ripple into *sustained*, positive change for our beautiful blue planet.

CHAPTER 4

From Suds to Sustainability: Rethinking Your Laundry Routine

Just like cutting down on laundry saves your clothes, easing up on our planet saves us all. Turn the tide like a spin cycle—clean clothes, conscience, clean planet.

In Chapter 3, we explored how to shop for better cleaning products for your home, particularly recognizing potentially harmful chemicals. The same ethical criteria that guide your general cleaning choices can also be applied to your laundry regimen. Remember the "ingredients to avoid" chart on page 69 for a helpful reminder.

Laundry detergent, a staple in our households, sees regular and substantial use. As a result, the market for liquid laundry detergent was valued at a staggering $27.1 billion in 2019.

So, why does this matter? When choosing between biodegradable laundry detergent and conventional variants, the former emerges as the clear victor for our environment. Choosing a biodegradable laundry detergent impacts our immediate surroundings and significantly reduces our ecological footprint.

However, understanding what distinguishes a biodegradable laundry detergent from a conventional one is essential.

Biodegradable laundry detergent comprises ingredients (notably surfactants) that can be broken down quickly when they find their way into our ecosystems. Then, microorganisms active in nature work on these compounds, helping to leave no trace of them behind.

Once they have been broken down, these ingredients do not go on to alter or harm the environment in any way. In Europe, detergent is considered biodegradable when all of the surfactants in the product degrade by more than 60 percent within twenty-eight days under aerobic conditions—in the presence of oxygen.

On the other hand, conventional laundry detergent is made of petroleum-based ingredients, dyes, synthetic fragrances, and other chemicals that do not fully degrade and go on to pollute our waterways and ecosystems.

Take petroleum distillate, for example. It is a surfactant that is commonly used in traditional detergents. Yet, it has been found to damage the mucous membranes and lungs, among other things.

Conventional laundry detergent can have a severe impact on the environment. It winds up in our ecosystems when wastewater from our laundry is discharged into sewers untreated. This has many consequences on aquatic life.

For starters, laundry detergent can cause eutrophication, where ingredients like phosphates and nitrogen (commonly used in conventional detergents) accumulate in rivers and other water bodies, causing algal blooms. When all of this algae dies, its decomposition sucks up all of the oxygen in the water, leaving other aquatic life without enough oxygen to survive, causing them to die too.

Where infiltration happens, the harmful ingredients found in conventional laundry detergent can also go on to contaminate groundwater. It leaves unwanted compounds like nitrates, ammonium, boron, and phosphates. This is very serious because groundwater is a significant source of fresh water for humans. In the US, groundwater supplies 51 percent of the population's drinking water.

POWDER VS. LIQUID DETERGENTS?

As we begin, consider this: on average, Americans complete approximately two and a half laundry loads per person weekly. This accounts for around 22 percent of your home's water usage. If you're not utilizing an appropriate detergent, you could be squandering even more water by excessively diluting liquids or necessitating additional rinse cycles to rid your clothes of powdered residues. Many detergent brands are available, but it's beneficial to consider the form your detergent takes—powder or liquid.

Powder Detergent

Pros: Elements such as bleaching agents and surfactants (the components that cleanse your clothes) maintain greater stability in powders. Hence, they possess an extended shelf life in comparison to liquids. Purchasing powders in large quantities—thus reducing surplus packaging—poses no concern about the detergents losing effectiveness over time.

Cons: Overuse results in a chalky white residue clinging to your clothes—necessitating further rinse cycles and, subsequently, more water. For some powders to dissolve fully, warm water is preferable, which could expend more energy than washing with cold water, a simple task with liquids.

Liquid Detergent

Pros: Liquids dissolve more efficiently in cold and warm water, eliminating concerns about residues on your garments.

Cons: The creation of liquid detergents requires water. Standard, nonconcentrated detergents may comprise up to 80 percent water. It's wasteful in terms of water and energy to transport diluted detergents nationwide when your washing machine can effectively convert powders into liquids using water from your local supply. Furthermore, a recent

Consumer Reports analysis highlighted that liquid laundry caps could lead to severe overuse, incurring additional costs and creating buildup in your machine. Finally, the measurement lines often need to be more adequately marked, making it difficult to determine the correct amount for a small, medium, or large laundry load.

Which to Choose?

When evaluating stain removal, powder detergents effectively remove specific stains like dirt, clay, and mud.

On the other hand, liquid detergents are superb at handling grease and body soil and offer a more convenient approach for pretreating stains.

A handy tip for those who prefer cooler wash temperatures: mixing powder detergent with a small amount of warm water before adding it to the washer could be beneficial to ensure it fully dissolves. This practice, however, may depend on your washing machine's specific design.

So, which is the better choice? Both liquid and powder detergents have unique benefits, and the choice between the two often comes down to personal preference. Choosing powders may be more environmentally friendly. You can quickly dilute them yourself and avoid powdery residues on your clothes by using less than you think is necessary. It's worth seeking out a brand specifically formulated for use with cold water. The provided scoops with powder detergents are easier to read, reducing the risk of using too much detergent.

Adding a half-cup of white vinegar to the rinse cycle can be beneficial. Powdery deposits sometimes arise from hard-water minerals that mix with detergents and redeposit on your clothes, not from the powders themselves. White vinegar helps remove these hard-water residues. Also, allow the powder to dissolve fully before adding your clothes to the wash, particularly for those with top-loading machines.

If you are firmly in the liquid detergent camp and can't bear to part with your plastic bottle, consider purchasing a triple-concentrated brand, such as Ethique. Alternatively, opt for a double-concentrated product over a full-strength version. These detergents use the least water and packaging, making them a greener choice.

Product Recommendations

Dirty Labs Bio Enzyme Laundry Detergent

The Unscented Co. Refill Laundry Detergent

Molly's Suds Powdered Detergent

Happi Earth

Elva's All-Natural Laundry Wash

Ethique's Zero Waste Laundry Detergent Bar

ARE LAUNDRY DETERGENT SHEETS BETTER FOR THE ENVIRONMENT?

It's complicated! Polyvinyl alcohol, also known as PVOH, PVA, or PVAL, is a synthetic polymer that's colorless and odorless. It's a chemical compound consisting of many molecules linked together.

PVA was discovered in the early 1920s and first commercialized by a Japanese chemical manufacturer who used it to package unit-dose pesticides. This protected farmers from chemical exposure, and the water-soluble pouches allowed the fertilizer to reach plants. Over time, these water-soluble properties have been utilized in many single-use, "almost" zero-waste products.

The manufacturing process of PVA begins with ethylene, which is also used for making other types of plastics like PET, high- and low-density polyethylene, and polystyrene. First, the ethylene is polymerized to become polyvinyl acetate (found in glue), then dissolved in ethanol or methanol with heat and a catalyst to produce polyvinyl alcohol.

PVA has multiple uses thanks to its strength, flexibility, and resistance to grease, solvents, and oil. It is also water-soluble and biodegradable, making it suitable for personal care, medical, household, and industrial applications.

PVA is typically considered non-toxic, with a low hazard score. In addition, it is recognized as a safe ingredient by the FDA and doesn't accumulate in the body, even when consumed.

However, despite being deemed safe, PVA's eco-friendliness is under scrutiny. Being derived from petrochemicals and technically plastic, it releases substances that can harm marine life. While it is commonly labeled as "biodegradable," a more accurate term would be "dissolvable." Under specific conditions and with certain microbes, PVA can break down from a polymer to a monomer. Yet, this process could take years, decades, or even centuries, posing a potential risk to marine life, similar to microplastics.

Moreover, while PVA is considered biodegradable, most wastewater treatment plants must be designed to degrade it, raising further environmental concerns effectively. For example, it's estimated that 75 percent of detergent pods containing PVA go untreated, potentially entering groundwater, ecosystems, and the human food supply chain. More than eight thousand tons of PVA residue is thought to be introduced into the environment annually in the US alone.

The current alternatives to PVA are limited. While the development of PVA from biobased sources is still in its early stages, we may see PVA made from plant starches, polysaccharides, or cellulose instead of fossil fuels in the future. Until then, it's crucial to consider the environmental implications of PVA while opting for more eco-friendly brands.

I should add that laundry detergent sheets are compact and lightweight, making them space-saving and travel-friendly. They reduce carbon emissions due to their smaller transportation footprint. The pre-measured sheets prevent detergent waste and dissolve fully in the wash, eliminating residue. They are typically free of harsh chemicals, providing a gentler alternative for your skin and clothes. Lastly, they can be used in all washing machines and for cold and hot water washes, increasing their versatility.

Product Recommendations

EcoRoots Laundry Detergent Sheets

TruEarth

Earth Breeze

Plantish Laundry Strips

DECODING FABRIC SOFTENER AND BLEACH

Fabric softener has existed for many years, gaining a foothold as a multi-billion-dollar industry in the name of floral fragrances and supposed "laundry care." So much so that in 2018 the global fabric softeners market size was at US$ 16.53 billion and is estimated to be around US$ 19.7 billion by 2026. But why is this product so popular in the first place?

To answer this question, we need to understand how fabric softeners work.

Approximately five main types of fabric softeners on the market can be classified according to the active component that interacts with the fabric to modify its properties: non-ionic, cationic, anionic, amphoteric, and reactive softeners.

However, most of the ones you find in your nearest grocery store act with a cationic surfactant principle; in other words, they adhere to the fabric's fibers to make it softer.

As clothes are washed and fabric softener is poured in, it adds a thin layer of chemicals designed to "lubricate" them, which prevents static cling. It makes the garments slippery to reduce friction. This means that once washed, clothes are less likely to wrinkle or accumulate lint, thanks to the anti-static nature of the blue liquid. Plus, it adds an extra pleasant scent.

But is fabric softener bad for your clothes? The dark side of fabric softeners starts with the fact that they aren't effective on all fabrics.

Yes, cationic surfactants improve the softness and smell of fabrics made from natural fibers with a high cellulose content, such as cotton, linen, or hemp. However, their effect is limited on other natural fibers such as wool or cashmere—almost nonexistent. The same thing happens with synthetic fibers such as nylon or polyester, so it's a waste of money to use softeners on them.

Talking about synthetic fibers, I hope you have kept your sportswear away from fabric softener at all costs, as the chemicals in fabric softener damage those clothes.

If we coat elastane or nylon with a waxy-like film, it's like clogging a filter and blocking its ability to release moisture. Of course, sportswear must remain breathable to prevent excessive sweating and unpleasant odors, but if that fundamental quality is removed, be prepared to welcome funky smells and excessive bacteria.

Furthermore, the myth that fabric softener "washes clothes better" is untrue and quite the opposite. The thin film of fabric softener may build up over time regardless of the type of fabric—including microfiber towels, making it difficult for water and detergent to penetrate it. And don't get me started on the hard-to-remove waxy stains that appear on clothes if you mistakenly poured more liquid than you should.

Most fabric softeners sell the softness + scent combo we all love, but at what cost?

One of the most questioned impacts of fabric softeners is their negative environmental impact. First, they usually come in plastic bottles, and it's no secret that plastic bottles are one of the biggest polluters worldwide due to their lack of biodegradability. Half of all plastic produced globally is made to be used only once and then thrown away, and let's be honest; only some people reuse their Downy bottles.

Besides generating plastic waste, fabric softeners contain several harmful substances, such as quaternary ammonium or "quats or QACs," non-biodegradable and toxic to aquatic microorganisms. Of course, it goes without saying that once these chemicals are washed down the drain, they cause an imbalance in the environment. But guess what? Mother Earth is not the only one affected by the soup of nasty chemicals generated by the laundry industry.

It turns out that some fabric softeners aren't even cruelty-free, as an ingredient derived from animal fat called dehydrogenated tallow dimethyl ammonium chloride can be found in them.

According to several studies compiled by the Environmental Working Group, both quats, fragrances—phthalates, ugh—preservatives, and dyes found in the blue liquid can trigger allergies, skin irritations, and even cause reproductive harm.

So far, I've only discussed liquid detergent, but these aren't the only crooks in the laundry department. So give a big boo to liquid detergent's cheap cousin: **dryer sheets**.

Those disposable single-use sheets you throw in your dryer contribute to environmental waste and have been found to emit more than twenty-five volatile compounds that pollute the air from dryer vents, potentially affecting human health and air quality. They are made from polyester, hello fossil fuels, and coated with quaternary ammonium compounds.

So, what are the alternatives? I have been recommending wool dryer balls for years. But, after reading a new study titled "Shear Destruction: Wool, Fashion and the Biodiversity Crisis," wool production might not be as eco-friendly as we believe. The report states that wool's environmental impact is immense, contributing significantly to climate change, biodiversity loss, and land and water pollution. Additionally, it uses a whopping 367 times more land than cotton. Despite these concerns, the wool industry is often marketed as sustainable, leading 87 percent of consumers to perceive wool as environmentally safe. Therefore, while wool dryer balls might seem natural and harmless, they are part of a much larger, complex problem, which we will discuss further in Chapter 5.

If you're searching for an alternative to wool dryer balls, ATTITUDE's Static Eliminator and Softener reusable dryer sheets are an alternative to wool dryer balls. They can be used for up to five hundred loads, reducing waste. However, be mindful that not all reusable dryer sheets are created equal—some may not be sustainably made or could contain chemicals. Also, note that many (including Attitudes) are made from polyester or nylon, which aren't biodegradable.

 DIY DELIGHT NATURAL VINEGAR DRYER SHEETS

Ingredients/Instructions:

- Cut up some old towels into 6-inch squares.
- Add ½ cup of vinegar and ½ cup of water to a large container; a wide-mouthed mason jar will work.
- Put the clothes in a jar to soak up the liquid.
- Toss 1 cloth into the dryer as needed.
- You can also use vinegar alone; add about ½ cup to your wash cycle.

Can you dry your clothes without dryer sheets? The answer is a resounding *yes*! Do not be afraid to ditch the sheets and go all natural.

Tips on removing static and scenting your laundry naturally:

- Dry synthetic fabrics like nylon and polyester separately because they cause the most static cling.

- Don't over-dry your garments; over-drying is a significant cause of static. You can help prevent this by putting your dryer on a low tumble.

- Humidity in the air helps to prevent static just as nicely; you can keep a humidifier in your laundry room, or if you are already hanging drying items, this will create it naturally.

Regarding hygiene, fabric softeners, and dryer sheets have nothing to do with making clothes cleaner. Initially, fabric softeners became popular because the detergents used at the time made clothes rough. Still, detergents have evolved, and many other ways of softening clothes are available that don't risk your well-being or the environment.

To revisit a key topic from Chapter 3, bleach, particularly typical household bleach, remains a glaring red flag in my perspective. Despite the assurances from manufacturers about its safety, it harbors significant concerns.

Foremost among these concerns is household bleach, an organochlorine compound. These compounds are notorious for their durability and resistance to degradation, which means they can persist in the environment for a long time, often accumulating in the food chain and posing severe ecological risks.

Additionally, the effectiveness of bleach in making clothes whiter has its fair share of controversy. While it does have the potential to brighten whites, excessive or improper use can cause yellowing or weakening of the fabric. Ultimately, the trade-off between momentary brightness and potential environmental harm raises severe doubts about its necessity in our households. So what can we use instead? Hydrogen peroxide is a mild yet effective bleaching agent for more than sun-kissed hair. This eco-friendly alternative to chlorine bleach is an excellent choice for stain removal and brightening your wardrobe. Upon exposure to light, it harmlessly decomposes into water and oxygen, acting as an oxygen-based, biodegradable bleach. To use it, add one cup of the standard 3 percent hydrogen peroxide solution, which you can find in the first aid section of your local pharmacy, to your washer with each load of laundry.

Whether for white or colored clothes, hydrogen peroxide works without damaging fabric color. However, ensure it's evenly dispersed in the washer with water before introducing your clothes, as pouring it directly onto dry, colored fabrics could lead to color loss.

Baking soda is another laundry booster you can consider. Half a cup combined with regular detergent enhances cleaning efficacy, leading to brighter, whiter clothes. Sprinkle it directly onto the washer drum before loading your laundry. It works perfectly in both standard and high-efficiency washers. If you can't completely let go of chlorine bleach, baking soda can reduce the amount you need and boost its cleaning power.

Oxygen-based bleach is yet another earth-conscious option that doubles as a potent whitening agent and stain remover. Gentle on both the environment and your fabrics, it is safe to use on all washable materials except silk, wool, and anything with leather trims. Unlike chlorine bleach, which strips color, oxygen bleach maintains fabric color while tackling stains and dullness.

For optimum results with oxygen bleach, presoak your laundry for at least two hours or overnight before your usual wash. Then, follow the instructions on the packaging for the correct amount per gallon of water. Brands like **OxiClean**, **Nellie's All-Natural Oxygen Brightener**, and **OxoBrite** offer powdered oxygen bleach, which tends to outperform their liquid counterparts regarding stability and efficacy.

If you're dealing with white fabrics, a soak in a solution of one part distilled white vinegar and six parts warm water overnight can work wonders. The vinegar smell will dissipate after washing, leaving you with bright, fresh linens and clothes.

Like vinegar, lemon juice is a natural bleach. Its acidic nature helps brighten white clothes, especially heavily stained white socks. Add one cup of lemon juice or sliced lemon to boiling water, turn off the heat, and let the items soak overnight.

Finally, for the ultimate eco-friendly solution, harness the power of the sun's ultraviolet rays. Besides saving energy and reducing your carbon footprint, sun-drying your clothes naturally whitens, disinfects, and fades stains. However, be mindful not to sun-dry colored fabrics, as the UV rays could cause fading.

STAIN TYPE	CLOTHING	RUGS/UPHOLSTERY	GRANITE/MARBLE
wine	rinse cold, soak in oxygen bleach 1-4 hrs, wash	dab with water/ bleach alternative, then hot water, blot	baking soda & peroxide paste, cover 24 hours, wipe, reseal
ink	rub alcohol & cotton swab, lift stain	rub alcohol, rinse, air dry	same as rugs, avoid drying alcohol on stone
grease	50% vinegar/ water, treat with laundry soap	dab with soap, rinse use vinegar if persists	
pet urine	wash with detergent hot, vinegar in rinse/soak	absorb, vinegar solution, air dry, baking soda & peroxide mix	
grass	stain remover/ laundry detergent, wash soak in bleach if persists		
coffee	hot water, 50% vinegar/water, laundry soap	blot, dish detergent & warm water, rinse, air dry, vacuum	baking soda & peroxide paste, cover 24 hrs, wipe, reseal
mud	laundry soap & brush on fabric	wet, soap & brush, vinegar spray blot	
blood	rinse cold, stain remover, wash soak in bleach if persists	cold water, dish detergent use ammonia if persists	

WHY I DON'T RECOMMEND DIY LAUNDRY DETERGENT

Cleaning, whether scrubbing your kitchen floor or washing your favorite shirt, is a dynamic process that hinges on a three-pronged energy approach: chemical, mechanical, and thermal.

First on the roster is chemical energy. Your laundry detergent or soap delivers this one. Here's the kicker: Detergents aren't just passive players in the cleaning game. They're like secret agents, armed with chemical energy that's attracted to the stains on your clothing. This energy bonds to the dirt and grime, ripping them from the fabric and into the wash water.

Next up is mechanical energy. Your washing machine is the star of this show, churning and spinning your clothes to pry the stains loose from the fabric. It's like a mechanical massage for your clothes, working the detergent deeper into the fabric. But washing machines aren't the only providers of mechanical energy; good old-fashioned handwashing can do the trick too!

Lastly, we have thermal energy, essentially the temperature of your wash water. Hotter water can dissolve stains quicker, aiding the detergent's cleaning process. While warmer water can enhance cleaning, today's laundry detergents are often designed to work well even at lower temperatures.

It's crucial to balance these three energy types. Depending on the task, they work together, but you can tweak their proportions. For example, delicate garments may require less mechanical and thermal energy, so we can lean on chemical energy and extend the washing time to ensure they get clean.

Now, let's address a contentious issue: DIY soaps or detergents. These homemade concoctions often include a mix of soap (either grated or melted), softeners, and boosters. Some recipes even exclude any actual cleaning agent! The problem with this is two-fold.

First, soap is not formulated for laundry cleaning. Soap is great for cleaning your skin, but when used on fabric, especially in a washing machine, it can leave behind a residue called soap scum. This residue can trap soil in your clothes over time, making them look less clean.

Second, the proportions in these homemade mixtures are often skewed, with a hefty helping of softeners and boosters but only a tiny bit of soap. This imbalance means you have plenty of agents to soften water and boost cleaning power but minimal actual cleaning agents to do the heavy lifting.

When using DIY detergents, there are two likely scenarios. One, you use a small recommended amount per load, and your clothes don't get clean because more cleaning agents must be needed. Or, two, you throw caution to the wind and use a large amount. While this could improve cleaning in the short term, the soap scum and trapped soil will likely start causing issues like reducing towel absorbency and making your clothes look dingy over time.

So why does soap differ from detergent? Yes, they both aim to remove soils and bacteria, but they're designed for different purposes and materials. Soap works well on non-porous surfaces that don't trap dirt, while detergent is made for porous, textured materials like fabric. It traps soil and suspends it in water, allowing it to be washed away.

Lastly, why are homemade soaps so prevalent and often recommended? They're seen as cost-effective and contain fewer chemicals. But remember, everything is made up of chemicals, even your homemade soap. It's essential to distinguish between harmful chemicals and those necessary for a product to function effectively.

WHAT ABOUT SOAP NUTS?

I tested soap nuts and saw if they could replace my laundry detergent. These little nuts, derived from the soapberry tree, contain a natural cleaning agent called saponin, which works wonders on clothes. Intrigued by their eco-friendly claims, I eagerly tossed them into the washing machine with heavily stained clothes.

The results were mixed, to my surprise. While some stains were successfully removed, notably lighter ones like dirt or food spills, tougher stains like oil or ink didn't budge much. It became apparent that soap nuts have their limitations regarding more stubborn stains.

That being said, soap nuts did excel in other areas. They effectively removed odors from my laundry, leaving them fresh and clean, albeit without a noticeable scent. This can be great for those who prefer unscented or are sensitive to fragrances. Unlike conventional detergents, I appreciated that they didn't leave any residue or buildup on my clothes.

I found soap nuts are particularly well-suited for lighter loads or delicates requiring a gentler touch. They can effectively clean and freshen these items without any issues. However, I learned it's best to turn to other stain-fighting methods or enzyme-based detergents for better results for those tough stains or heavily soiled garments. Soap nuts offer a natural and eco-friendly alternative for laundry, but it's essential to understand their limitations and adjust expectations accordingly. They have strengths in odor removal and gentler cleaning, but it's best to have a backup plan for challenging stains.

CLEAN CLOTHES, DIRTY OCEANS

It's no secret that plastic is a significant player in our everyday lives, but did you know that microplastics—tiny fragments of plastic, generally five millimeters or less in size—are a growing concern for our planet? These microplastics come from various sources, including larger plastic items like bottles that break down into smaller fragments, car tires, plastic beads in skincare products, and synthetic fibers. It's fascinating to think about how quickly our understanding of environmental issues can change. Just a decade ago, scientists were only beginning to explore the impact of microfiber pollution on our oceans. Then, a groundbreaking study looked at shorelines across six continents and found that laundry was a significant source of plastic pollution in the world's oceans.

About 35 percent of the tiny plastic bits in our oceans likely come from washing clothes, sheets, and towels. The washing process creates friction that leads to the release of tiny plastic fibers, known as microfibers. These typically enter our environment through wastewater.

Textiles are now considered the largest known source of marine microplastic pollution, with approximately 2.2 million tons of microfibers entering the sea each year.

The way the yarns are twisted together plays a significant role in this. When we wash our clothes, the water, friction, and detergents can cause those filaments to shed. So, if you have a piece of clothing with tightly twisted yarns and a tight weave, it will shed less than a piece with loosely twisted yarns and a loose weave. It's all about the structure of the fabric!

Recently, a white paper to the European Commission suggested that washing machine filters might be the key solution in tackling the issue of microfiber pollution. These nifty devices have the potential to block more than 90 percent of microfibers from entering our water system.

There needs to be more debate over how efficient these filters are. Some suggest they can block anywhere from 29 percent to 74 percent of microfibers. Other strategies are still being explored, like developing fabrics that shed less or trapping fibers at wastewater treatment plants. But, for now, washing machine filters are a good starting point.

This has led to legislation in various countries. For example, France has already decreed that by 2025, all new washing machines need to be able to filter microplastics. Australia is also moving in this direction, with plans to introduce microfiber filters in all new washing machines by 2030 as part of their national plastics plan.

However, these filters are not exactly pocket-friendly. Retrofitting your washing machine with a microfiber filter could set you back anywhere from one to three hundred dollars, depending on the brand. And that's not considering the space you'll need. In addition, these filters can be quite large and need to be cleaned or replaced every twenty washes. So, they must be installed somewhere convenient and budgeted for regular maintenance.

Once installed, correctly disposing of the trapped microfibers is another critical step. Some filters let you brush the fibers into the trash, while others use a return-and-replace system. Companies like PlanetCare are even thinking about recycling these microfibers into insulation mats.

If you're short on space or find the filters too expensive, the **Guppyfriend** bag is another way to help reduce microfiber pollution.

Just pop your synthetic clothes into the bag before washing, and it will trap the loose microfibers. Using washing machine filters could prevent 90 percent of the thirty-five tonnes of microplastics discharged daily in the EU.

But while these filters and bags can make a difference, it's crucial to remember that solving the microfiber problem requires more than individual actions. It's a collective effort that includes industry changes and legislation.

Until then, there are other things we can do in our daily lives to lessen the issue:

- Wash clothes less often
- Wash inside-out
- Store properly
- Repair damage
- Wash on full loads
- Avoid using the delicate setting on your washing machine
- Consider using a front-loading washing machine, which tends to cause less shedding
- Whenever possible, line-dry your clothes

Let's get straight to the point—just because it's cold outside doesn't mean you have to abandon your clothesline. You'll need heat, humidity, and time to dry your clothes effectively. During winter, you'll have to compensate for the lack of heat by focusing on the other two factors. If there's any breeze, hang your clothes outside as early as possible to take advantage of the entire day. However, it's best to dry indoors on grey and damp days. Don't worry about snowy and cold days; your clothes might freeze but eventually dry out as the ice becomes vapor. Windy and cold days might make your clothes stiff as ice sculptures, but the breeze will help soften and dry them.

Frozen clothes might seem daunting, but a quick spin in the dryer can easily fix them once you bring them inside. On sunny days, hang your white clothes outside to take advantage of the sun's natural bleaching power. If it's not raining, any clothes will do.

💡 Keep a clothesline from obstructing your backyard view or sunny spots. Instead, use a retractable line that can be put away when unused. If you're a DIY enthusiast, you can rig up your clothesline using sturdy materials that won't sag or rot over time. Another option is a rotary clothesline, which may not be the prettiest thing but is incredibly practical as it turns the clothes for better air circulation. According to Project Laundry List, you can save more than twenty-five dollars per month on energy costs by avoiding the dryer.

> **TIPS FOR AIR-DRYING YOUR CLOTHES**
>
> - If drying jeans or pants, fold the waistline over the line and click with a clothespin. Opt for wooden clothespins, not plastic, and don't leave them outside, like the DIY line. They can rot.
> - Before you hang anything, give the line a good wipe, especially if it's outside.
> - Hang all other items from the hem to avoid a dent from the clothespin.
> - If you want to avoid ironing, bring the clothes in as soon as they are dry, smooth them out with your hands, and that's it. You're done!

Laundry has a considerable environmental footprint that we often overlook. Think about it: the average American family does around three hundred laundry loads each year, using about 12,000 gallons of water and consuming vast amounts of energy. The carbon dioxide emissions from doing laundry are equivalent to the annual emissions from around 39 billion cars! That's a staggering number, considering only about 1.4 billion cars are on the road globally.

But don't worry. There are numerous ways to lessen this impact. One significant game-changer is replacing your old washing and drying machines with ENERGY STAR–certified appliances. These devices use roughly 25 percent less energy and 33 percent less water than their regular counterparts. If every American household made this switch, we could save over $3.3 billion yearly and prevent a whopping 19 billion pounds of greenhouse gas emissions! Equivalent to removing about 1.9 million cars from the road for one year.

When it comes to washing, cold is the new hot. Most of the energy a washing machine uses goes toward heating water. The best part is colder temperatures are also easier on your fabrics, reducing wear and tear and preventing the release of up to 700,000 tiny synthetic fibers that could pollute our water bodies.

Hot water may be necessary for cleaning bed linens after someone's been sick or dealing with your sweaty gym clothes. But aside from these specific circumstances, cold water should do the trick.

Drying your clothes also consumes a lot of energy, with dryers often using five to ten times more power than washing machines. However, newer technologies are helping reduce this impact. For instance, heat-pump dryers recycle hot air, making them more energy efficient.

HOW TO CHOOSE AN ENERGY-EFFICIENT WASHING MACHINE

Saving energy and water by switching to a sustainable washing machine is a win for the environment and your pocketbook! This is because 90 percent of all energy used to wash your clothes comes from just waiting for the water to heat up. With the average household doing almost three hundred loads of laundry a year, this is where energy-efficiency washers (with cold-water settings, smaller drums, and shorter cycles) can make a huge difference.

While eco-friendly washing machines can sometimes be more expensive, you will save money in the long run! If you buy a high-quality one, it should last seven to ten years. That is a lot of years to cut down your energy and water bills!

Lastly, eco-friendly machines are generally much quieter and gentler on your clothes. Traditional washers are loaded from the top and use an agitator to free dirt and oils. While effective, agitators can be noisy and rough on delicate fabrics. Most eco-friendly models are front-loading washers since they use about 45 percent less energy and 50 percent less water.

Things to Consider When Shopping for an Eco-Friendly Washing Machine

DOES IT SAVE WATER AND ENERGY?

This one might seem slightly obvious, but the energy and water usage of different washing machines marketed as "eco-friendly" vary! This is why making an informed purchase and researching is essential before buying.

ENERGY STAR will be your best friend here since ENERGY STAR–certified washers use less water and energy to run. They do this by identifying the Integrated Modified Energy Factor (IMEF) and Integrated Water Factor (IWF) of a model. IMEF measures the energy used to run the spin cycle and heat the water. While IWF estimates the gallons of water consumed per cubic foot of drum capacity. A water-efficient washing machine will have a low IWF, while an energy-efficient model will have a higher IMEF. Their homepage has an easy selection process where you can filter out different brands, price ranges, and more! You can also check their yellow EnergyGuide label to estimate the model's energy use and compare it to similar models.

IS IT SECONDHAND?

A secondhand washing machine? "Candice?" you ask me. Yep! Buying a model that is energy efficient and secondhand gives you the power and water-saving benefits down the road, but you are also offsetting the footprint you would have gained from manufacturing and transporting a new machine.

DOES IT HAVE CERTIFICATIONS TO BACK IT UP?

As always, you want to be on the lookout for greenwashing companies and have the certifications to back up their eco-friendly claims (this goes for dryers, too, by the way). When buying an eco-friendly washer and dryer, consider certifications like ENERGY STAR. AAFA-certified washers also use steam to remove bacteria and dust mites from your clothing; this can be an excellent alternative for anyone with allergies and sensitivities to the toxins in laundry detergent.

WHAT DO REVIEWERS SAY?

I always check the reviews before I buy anything these days. And when doing so, you must consider what is important to you! For example, when reviewing reviews of front-load washers (most eco-friendly washers are front-load), one of the most common complaints is that they tend to develop mildew and a smell faster than top-load agitator washers. While annoying, you can always use a green cleaner to clean your front-load washer and combat smells!

DOES THE COMPANY HAVE TRANSPARENT REPORTING?

Be on the lookout if the company you buy from has transparent reporting for their scope 1, 2, and 3 emissions! LG and Samsung are two examples of companies that sell efficient washing machines and are also on the reputable side for reporting their carbon emissions.

IS IT TOO BIG OR TOO SMALL?

Every eco-washing machine has a different width and drum capacity; larger drums use more water and energy with each load. If you have a large family and do laundry quite often, you might opt for a larger size. But stick to a smaller size if you are just one person, and you must fill up your washer to the top before running it.

AM I BUYING TO BUY?

As I mentioned, a considerable portion of washing machines' environmental impact comes from the unit's manufacturing and delivery. Remember, the second R in the nine R's of Zero Waste is *Refuse*. This means that even if you buy a new efficient washing machine, the energy and water savings you will gain might not make up for the emissions produced to get it to your home in the first place. So with that in mind, only buy a new machine if your old one is beyond repair!

WILL IT NEED REPLACING SOON?

I recommend checking out the manufacturer's repair and warranty policies before you purchase an eco-washing machine. From an environmental standpoint, repairing something broken is always better than buying new ones! For context, a good-quality washing machine should last between seven and ten years.

WHAT ABOUT DRYERS?

When shopping for an eco-friendly dryer, remember that eighty-eight million dryers in the US alone emit over a ton of carbon dioxide annually, equivalent to approximately the emissions produced by driving a car for around 4,800 miles.

Here are some things to consider:

1. **Certification Matters:** Always go for dryers with an ENERGY STAR certification. This label ensures that the appliance has met strict energy-efficiency criteria.

2. **Size Does Matter:** A larger-capacity drum would be ideal if you frequently wash and dry more oversized items or do lots of laundry. It allows clothes enough room to tumble around, leading to efficient drying. Conversely, a compact or portable machine could be your ticket to eco-friendliness if your laundry needs are modest.

3. **To Vent or Not to Vent?:** If you have an existing vent for your dryer, keeping it vented provides more options. However, if you don't, ventless dryer options are available. These machines condense the moisture and require the water to be manually removed or sent to a drain.

4. **Plug Play:** You may need a special plug for electric dryers. However, if you're replacing an old dryer, you can use the existing plug.

5. **Gas vs. Electric:** Methane, or "natural gas," is a potent greenhouse gas, a compelling reason to choose an electric dryer over a gas one. Electric dryers are becoming faster and more efficient, narrowing the speed gap between them and their gas counterparts. Compact, ventless electric dryers can now dry towels in about ninety minutes. Note that tight gas dryers aren't widely available—they typically come in full size only.

Let's chat about the best practices for using your dryer! I know this might sound radical, but one surefire way to be super energy efficient is not to use your dryer. Yep, it's true! It's like energy efficiency on steroids!

Now, let's address the lint filter situation. After every load, give that trusty lint filter a good cleaning. Don't let it get clogged up like a traffic jam during rush hour. When the air circulation is blocked, your dryer's energy efficiency takes a nosedive. Plus, it's a crucial home safety measure. We don't want any lint-related mishaps, do we?

Here's a nifty feature for dryer shopping: a moisture sensor option. The cool kids on the block, a.k.a. newer dryers, come with this magic feature. It's like having a mind-reading superhero in your laundry room. It automatically shuts off the dryer when your clothes are dry, saving energy and sparing you from unnecessary wear and tear. It's a win-win situation.

Speaking of loads, find that sweet spot between "too full" and "barely there." You want enough clothes for a full load without blocking the air circulation. And here's a pro tip: avoid mixing wet clothes with partially dry ones. That's like throwing a wrench into the drying process and increasing the drying time.

When it's time to play wardrobe matchmaker in your dryer, sort your loads by the thickness of your clothes. Then, lightweights can have their little party in one quick load while the heavy hitters get their gig. It's all about creating a harmonious drying experience for each fabric type.

Here's a little secret: run your loads back to back. Why waste all that hot dryer goodness? Instead, take advantage of the already-toasty machine and immediately throw in your next load. Efficiency at its finest, my friend!

> **BONUS STEPS FOR A GREENER WASH**
>
> 1. Sort laundry based on fabric types, not just colors, to preserve shape and color.
> 2. Familiarize yourself with clothing tag symbols for optimal washing and drying practices.
> 3. Avoid using the dryer for fabrics like linen and silk to prevent damage.
> 4. Retire the iron to save energy (ironing consumes 1,800 kW of electricity).
> 5. Try line-drying clothes or removing them from the dryer while hot to minimize wrinkles.
> 6. Choose sustainable fabrics that naturally resist wrinkling for low-maintenance options.
> 7. Regularly service your washing machine and dryer to improve energy efficiency.
> 8. Be cautious when selecting a dry cleaner, opting for nontoxic alternatives like "wet cleaning" or silicone-based cleaning. Avoid dry cleaners using PERC (Perchloroethylene), a harmful substance.
> 9. Save on energy costs by adjusting your laundry schedule. Aim to wash before four o'clock in the afternoon or after seven in the evening to avoid peak hours when energy rates are higher. In the summertime, consider running your washer in the early morning when energy usage peaks during hot afternoons.
> 10. Use refills and concentrated detergents to reduce packaging waste and minimize chemical use.

Even the simple act of doing laundry can make a difference regarding sustainability. *Sustained* living means finding enduring and balanced ways to carry out our daily activities while considering the environmental impact. Sustainable laundry practices all contribute to the long-term well-being of our ecosystems.

CHAPTER 5

Sewing Seeds of Change: Your Guide to Ethical Fashion

Remember your favorite tee or pair of jeans from a few years ago? The one that wore out way too quickly? That's the result of "fast fashion"—a world where clothes come and go like passing trends, made quickly and replaced even faster. As we wade through aisles of temporary styles, it's hard not to miss the days when fashion meant pieces curated with care, intended to last beyond a season. This shift isn't just about clothes; it speaks to a more profound change in how we value, or perhaps undervalue, longevity in our wardrobe choices.

Sustainable fashion signifies an ethical approach that encompasses the design, sourcing, production, sales, and distribution stages of clothing, aiming to reduce adverse effects on people and the environment throughout its entire lifecycle.

So, how does one practically incorporate sustainable and ethical fashion into their life and become a more mindful consumer?

In this chapter, I'll guide you on adopting this approach while shopping: from scrutinizing raw materials to understanding the implemented procedures across the supply chain, all the way to considering a garment's disposal.

In today's paradoxical world, it's no surprise that a movie ticket might set you back more than a pair of jeans. This, my friends, results from the perilous phenomenon known as fast fashion. Famous brands like H&M, Shein, Forever 21, Target, Zara, and Primark are the usual suspects here, who have sped up fashion cycles, transforming the classic four seasons into a whirlwind of fifty-two micro-seasons—an almost weekly wardrobe transformation. Consequently, trends fade as rapidly as they emerge. Even more extreme is the case of Shein, which is churning out, mind-bogglingly, over nine thousand new styles daily on its website, leading to a new term in the fashion world: "ultra-fast fashion." Given their disposable nature, it's hardly a shock when these garments show signs of wear after a single use.

While curtailing consumption is reasonable, modest shopping habits alone can't sweep away the murky underbelly of the fashion world that lies hidden beneath the glitz and glamour.

Let's delve into the significant challenges the industry must overcome to underscore the urgency of infusing sustainability into fashion.

1: CLIMATE CRISIS AND CONSUMPTION OVERDRIVE

Undoubtedly, the fashion industry is a leading player in exacerbating climate change, responsible for 8.1 percent of the world's total carbon emissions. One key offender is cotton production, with global cultivation churning a staggering 220 million tons of CO_2 annually, roughly equivalent to the carbon emissions produced by nearly forty-seven million cars in a year. The issue is amplified by the sector's reliance on fossil fuels, fueling the production, distribution, and manufacturing of petroleum-derived fabrics like nylon, spandex, and polyester, further escalating the carbon dilemma. Inefficient recycling techniques and poor waste management strategies add insult to injury, culminating in a heightened emission of greenhouse gases.

Even so-called green alternatives like organic cotton can unintentionally contribute to the problem. If discarded into landfills, these materials decompose in an oxygen-free environment, generating methane, which is twenty-five times more potent a greenhouse gas than CO_2, intensifying the climate change threat.

Fueling this environmental predicament is the rise of overconsumption. The world now consumes a staggering eighty billion pieces of clothing each year, which has risen by 400 percent compared to two decades ago. Each of us buys around eighty pounds of clothing annually. Furthermore, today's consumers purchase 60 percent more clothing items than just twenty years ago, feeding into the demand that exacerbates the industry's destructive impact.

However, amidst this dire situation, there lies a beacon of hope—*sustained* fashion. This approach includes using recycled or upcycled materials, powering factories with renewable energy, and implementing carbon offsetting practices to neutralize shipping emissions. By addressing production and consumption, sustained fashion could redefine the industry's relationship with our planet.

2: HUMAN RIGHTS ISSUES

With a workforce exceeding 300 million, the fashion industry has become a hotbed for numerous human rights violations. Hazardous working conditions are distressingly common, exposing countless workers to potential harm daily. Coupled with this are the mandatory long hours, and even overtime, which take a severe toll on the health and well-being of the workers. The specter of harassment, whether verbal, sexual, or physical, also looms large within the industry. These actions are detrimental to the overall working environment and seriously affect the mental health of the individuals subjected to it. A disturbing pattern of gender discrimination has been observed, primarily directed toward young female garment workers, further perpetuating inequality and injustice.

Moreover, the rampant exploitation of migrant workers is a stark reflection of the lack of ethical considerations within the industry. In a world where every individual should be guaranteed their fundamental rights, modern slavery practices within the industry is a deplorable reality. Lastly, many workers within the fashion industry receive woefully inadequate wages. These meager earnings are far from enough to sustain a decent standard of living, often forcing workers into a cycle of poverty and exploitation.

The 2013 Rana Plaza garment factory collapse in Bangladesh was a horrifying wake-up call. The disaster, which claimed the lives of 1,135 workers and injured a further 2,500, sparked global outcry and prompted probing questions about the hidden cost of cheap fashion. This pivotal moment gave birth to the movement known as the Fashion Revolution. Sadly, these human rights violations are deeply embedded at every stage of the supply chain: In the sourcing phase, farmers and processors are often subjected to unjust labor practices and exposed to harmful chemical pesticides and plasticizers.

In the manufacturing phase, workers' rights are virtually nonexistent in countries where unions are absent or lack power.

On the consumption front, the industry has been criticized for lacking diversity and inclusivity, promoting an unhealthy body image, and marginalizing minority groups.

All these violations reflect the pressing need to bring about a human-rights-centered transformation in the fashion industry.

3: SUPPLY CHAINS & TRANSPARENCY

Recently, there's been a roaring demand for transparency in the fashion industry. From consumers to advocacy groups, everyone wants to know the journey of their apparel from start to finish. But here's the deal: it's tough. Despite the growing calls, the 2022 Fashion Transparency Index shows that big brands are still pretty tight-lipped about their operations.

Why? Well, the fashion world is complex. Every clothing piece has a story, with multiple actors involved. Brands worry about their image—if they dig too deep, they might find things they'd rather not know. Plus, keeping an eye on every step is costly, especially when you have suppliers dotted all over the globe.

This brings us to ethical sourcing—essentially, it's about ensuring that what you buy hasn't harmed people or the planet. It provides workers fair pay, work in safe conditions, and the environment isn't being trashed in the process. Companies prioritizing this often have shorter supply chains—which are easier to manage and more transparent. These businesses are in tune with where they source from and are sharp on local labor laws. Regular checks on their suppliers are part and parcel of their operations. Being open about sourcing builds trust with consumers who know where their clothes come from. It's a win-win for brands: they stand out in a crowded market and vibe with consumer values.

Tech's also stepping up to help with the transparency puzzle. Take FibreTrace® MAPPED—it's like a digital detective for textiles, using blockchain to track a product's life story. If tools like this become mainstream, we could see a game-changing shift in fashion transparency. But tech alone isn't the answer. True transparency needs a deep-rooted commitment to being open and responsible throughout the supply chain.

Now, let's break down how a garment comes to life. It starts with raw materials—consider where and how cotton is grown. This then turns into fiber, which is spun into fabric. After some tweaks like dyeing, it's design time. The production involves everything from buttons to threads. Once it's ready, it gets some final touches—like tags—and off it goes to stores and customers. This process is a complex dance, changing with every season and every garment. And given this maze, even the most eco-friendly brands can unknowingly trip up. The challenge? Keeping tabs on every step and ensuring it's all above board.

4: TOXIC LOVE AFFAIR

The unyielding surge of chemicals in the fashion industry is a complex issue that has recently been thrust into the spotlight, revealing alarming facts and implications for environmental health and human safety.

Recent investigations by organizations such as the Center for Environmental Health in California and the Canadian Broadcasting Corporation have exposed high levels of toxic chemicals, from the hormone-disrupting BPA to lead and PFAS, in garments from popular brands including Nike, Athleta, Hanes, and even fast-fashion giants like

Zaful and Shein. These discoveries further the concerns about the fashion industry's growing ecological footprint, with substances like PFAS, a group of "forever chemicals," becoming synonymous with sustainability risks. These synthetic chemicals, pervasive in various daily items, from nonstick pans to clothing, linger in the environment and our bodies, not breaking down and potentially causing harm.

While some fabrics, like polyester, might inherently have a higher toxic profile than others, like organic cotton, the treatment these fabrics undergo poses a greater risk. An organic cotton jacket might begin its journey without toxic pesticides, but a finish laden with PFAS for water resistance can transform its ecological profile.

The chemical footprint of clothing doesn't solely originate from intentional additives like dyes or water repellents. Contaminants, unintended by-products, or residues also plague garments. For instance, lead might be present in a cotton t-shirt due to contaminated soil or as a color stabilizer in dyes. While PFAS may be added intentionally for specific attributes, contamination from machinery is also a source. With recent legislation targeting intentional addition, this concern is bound to be mitigated. However, intentional additives remain the dominant worry, given their concentration and sheer volume.

The dangers of these chemicals are manifold. Products like sports bras and undergarments that remain in close contact with our skin have been found with alarming levels of substances like PFAS and BPA. The latter, a hormone-disrupting chemical, is particularly concerning due to its potential effects on metabolism, growth, development, and reproduction. The presence of these chemicals in items intimately connected to our bodies elevates the potential for harm as our skin can absorb these toxins.

Highlighting the gravity of the issue, brands like Thinx and Lululemon have faced backlash and lawsuits due to chemical contaminants. Consumers are increasingly wary of the revelation that as much as 72 percent of "water-resistant" or "stain-resistant" products test positive for PFAS.

However, the fashion industry's path to redemption appears in pockets. Various states, including California and New York, have initiated bans on using PFAS in clothing and other items. Europe, too, is joining the crusade with proposals to ban these chemicals extensively. These measures, combined with consumer awareness and the push toward sustainable fashion practices, hope to weave a future where clothing is not only a statement of style but also a testament to environmental and personal health.

5: WATER WOES

The fashion industry's seemingly insatiable thirst for water is staggering, accounting for its position as the second-largest consumer of water globally. It guzzles a mind-boggling one to three trillion gallons of water annually. This intensive water usage paints a grim picture, considering over two billion people in more than forty countries are already grappling with water shortages.

One primary culprit of this excessive water usage is the conventional cultivation of cotton. Known as a "thirsty crop," its extensive cultivation has led to a 15 percent shrinkage of the Aral Sea in Central Asia, a poignant example of the fashion industry's environmental toll.

But the water demand doesn't end with cultivation. The production of a single pair of jeans can drain almost two thousand gallons of water, as water is an essential part of every stage in a garment's lifecycle. From the growth of fibers to the disposal of clothes, the industry has significant water-related environmental impacts, including high water usage, intense chemical pollution, and abundant physical microfiber pollution.

For instance, consider the water-intensive "wet-processing" stage, where raw materials are converted into textiles. This process employs over eight thousand synthetic chemicals, including chemical dyes that account for an estimated 20 percent of all global water pollution. The situation is dire in countries like China, where approximately 70 percent of freshwater is contaminated due to the fashion industry's 2.5 billion gallons of wastewater, as highlighted in the 2016 documentary *RiverBlue*.

The repercussions of this water use extend beyond the environment, affecting local communities as well. For example, the water used for irrigating cotton often depletes precious local aquifers and groundwater stores, impacting access to drinking water.

The fashion industry's high water consumption extends to something as basic as a cotton shirt, whose production alone can use up to 713 gallons of water. To put this in perspective, that's the same amount of water an average person would drink over two and a half years.

Given these alarming facts, it becomes evident that the fashion industry's environmental footprint extends far beyond mere carbon emissions. Its unchecked water usage and resultant pollution underscore the urgent need for sustainable practices and innovations to quench this thirst without draining the planet's resources.

6: MOUNTAINS OF TEXTILE WASTE

Fast fashion presents a significant obstacle to achieving a circular economy, notorious for generating a deluge of textile waste even before garments reach customers. From production offcuts to unsold "deadstock" garments, most of this waste is typically destined for incineration. In the US, a mere 15 percent of unwanted clothes find their way to recycling or donation facilities, culminating in a staggering twenty-one billion tons of textile waste ending in landfills yearly. Since 64 percent of these textiles are plastic-based, they persist in the environment without fully decomposing. Among the donated clothing, less than one-fifth is resold, with the majority exported to profit-driven textile recycling companies abroad. In contrast, sustainable fashion brands advocate for a circular economy, providing options for reselling or recycling old clothes regardless of the brand.

The stark reality is that in 2015, an alarming ninety-two million tonnes of clothing were discarded in landfills. If unchecked, the current growth rates suggest this figure could escalate by 50 percent by 2030. Astonishingly, of the 73 percent of garments consigned to landfills or incineration annually, an estimated 95 percent could have been reused or recycled. This mismanagement of resources translates to 87 percent of total fiber input used for clothing being landfilled or incinerated, leading to a missed economic opportunity exceeding $100 billion annually. Moreover, each kilogram of natural textiles in landfills releases four kilograms of CO_2, aggravating the climate crisis. A sustainable solution urgently needs to be implemented.

7: FROM FASHION RUNWAYS TO ANIMAL RIGHTS

Beneath the surface of the fashion industry's allure, a grim reality lies concealed: it thrives on a cycle of exploitation and slaughter of myriad animal species, leading to dire consequences. The industry often capitalizes on fur-bearing animals, including raccoons, dogs, minks, foxes, and coyotes, subjecting them to intolerable conditions in factory farm cages or violent ends in their natural habitats.

Exoticism in fashion sees reptiles like alligators, crocodiles, snakes, and lizards become targets, with their skins turned into coveted accessories for luxury brands such as Hermès and Louis Vuitton. Australia's native saltwater crocodiles, integral to the history of the Larrakia people, are especially victimized. Likewise, birds like ostriches and kangaroos are exploited for their skins to produce bags, boots, and football shoes.

The fashion industry also indirectly affects animals integral to the meat and dairy supply chains, such as cattle. Although cattle skins are viewed as by-products of these industries, the substantial economic value derived from their transformation into leather products perpetuates the cycle of exploitation.

Often, brands position their leather goods as "ethically sourced," indicating they are merely repurposing by-products of the meat industry. However, there's a semantic twist. A by-product, initially an incidental outcome of manufacture, becomes a profitable co-product once it acquires market value. For instance, when transformed into a wallet or jacket, the skin of a cow fetches a handsome profit, indirectly supporting the slaughter of animals. This financial symbiosis of the fashion and meat industries is concerning. The leather industry, evaluated at approximately $128 billion, indirectly bolsters the meat industry. Ethical alternatives, such as vintage, biobased leather, plant-based wool, and recycled-down, can counteract this systemic exploitation. Vegan leather is another option, but there are certain things I want you to keep in mind. Vegan leather, or faux leather or pleather, is typically produced from polyvinyl chloride (PVC) or polyurethane (PU). It presents a unique blend of benefits and challenges regarding sustainability.

On the upside, vegan leather eliminates the need for livestock, thus reducing deforestation, excessive agricultural land use, and animal deaths. It also has a lower carbon footprint than traditional leather, emitting 15.8 kg per square meter compared to leather's 17.0. In some instances, vegan leather is created from recycled plastic or natural materials like pineapple leaves, apple peels, or cork, bypassing plastic entirely.

However, vegan leather production is not without its pitfalls. The process can be chemically intensive, especially with PVC, which is highly polluting and can release harmful dioxins. Although PU is a slightly better alternative, it's still derived from fossil fuels. Vegan leather, especially synthetic variants, isn't biodegradable and can release harmful substances during decomposition. Moreover, it is often less durable than real leather, potentially leading to more waste due to frequent replacements.

Regarding environmental impact, it's a complex decision when comparing vegan leather to traditional leather. It depends on various factors, such as the production process, usage, and personal priorities. For instance, buying a long-lasting, eco-friendly real leather piece might have a lower overall environmental impact than replacing faux leather items frequently.

If sustainability is your main concern, you might lean toward high-quality real leather that lasts for years and is a by-product of other industries, preferably processed using vegetable tanning. PU is generally a better option for vegan leather than PVC, and natural materials are even more favorable.

However, from an ethical perspective, particularly for vegetarians and vegans, vegan leather remains a more coherent choice, negating the need for any animal involvement in its production. Therefore, the final decision concerns the aspect of sustainability most vital to you.

Farms, far from havens of pastoral serenity, serve as battlegrounds for animal rights. The livestock, whether grazing in picturesque pastures or confined to oppressive factory farms, endure grim living conditions and painful procedures. Many countries' legislations inadequately protect these animals, often legalizing cruelty if necessary or standard. Moreover, the fashion industry's production processes are causing biodiversity loss. As the production of certain materials demands extensive land use, many wild animals not directly killed for fashion suffer from destroying their habitats. Farmed animal-derived materials are usually the most land-intensive and, therefore, the most harmful to biodiversity. For instance, producing a single bale of wool necessitates 367 times more land than growing an equivalent quantity of cotton or hemp.

As awareness of these realities rises, change is on the horizon. The push for cruelty-free fashion and demand for legal animal protection signal a shift toward acknowledging animal sentience and safeguarding their rights. Our fashion choices can uphold these systemic issues or catalyze an industry-wide change toward a more sustainable and ethical future. Hence, the importance of discerning where our "dollar vote" goes cannot be overstated.

8: THE GREENWASHING MIRAGE

Greenwashing is a deceptive marketing tactic where a company claims to be more environmentally friendly than it truly is. It's used to appeal to increasingly eco-conscious consumers. The fashion industry, unfortunately, has its fair share of greenwashing examples.

While H&M has made strides toward sustainability with its Conscious Collection, which claims to use sustainable materials like organic cotton and recycled polyester, the company has also faced criticism. Critics argue that despite the Conscious Collection, the bulk of H&M's business still relies on a fast-fashion model, which inherently promotes overconsumption and results in vast amounts of waste.

The fast-fashion Chinese retailer Shein has appointed a head of environmental, social, and governance (ESG). This move aligns with current trends in the fashion industry, where sustainability is a growing concern, especially among Gen Z consumers. However, the decision might not be as impactful as it appears, given Shein's business model, which is often criticized for labor and environmental exploitation.

While Shein's initiative follows similar actions by other fast-fashion giants such as Zara, ASOS, Primark, and Boohoo, these brands' sustainability commitments have also drawn scrutiny. For instance, Zara has vowed to use recycled or sustainable materials only by an unspecified future year. Likewise, ASOS and Primark have issued similar promises.

Boohoo, recently under fire for poor working conditions and low wages at its Leicester factories, responded by publishing its supplier list and recruiting an ESG professional to its board last October.

Yet, these measures only sometimes equate to substantial change. Despite these steps, the pressure remains on brands like Shein to demonstrate a genuine and significant commitment to sustainability through appointing roles and lofty pledges and a thorough transformation of their business practices.

Many brands now offer lines made from recycled materials, often implying significant environmental benefits. However, the recycling process can be energy-intensive and polluting, depending on the technology used. Furthermore, there's usually a lack of transparency about how much of the final product is made from recycled materials.

Some brands publish sustainability commitments with ambiguous timelines and little detail on how they will achieve these goals. Without specific targets and regular updates on progress, these commitments can come off as greenwashing.

The fashion industry is notorious for its vague use of terms like "sustainable," "ethical," "green," and "eco-friendly" without offering clear definitions or proof to back up such claims. Without industry-wide standards, these terms can be used to greenwash products.

Some companies claim they are "carbon neutral," but often, this is achieved through carbon offsetting, such as investing in reforestation projects, rather than through actual reductions in their emissions. While carbon offsetting can be part of the solution, it can also be used as a greenwashing tactic if companies use it to avoid making substantive changes to their operations.

While sustainable brands often use third-party certifications and auditing schemes to ensure authenticity, even these are not foolproof. Renowned certifications like B Corp have had their share of greenwashing accusations, reminding us to maintain a healthy skepticism.

So, what's our next step? That is the profound question we face. *Sustained* aims to guide you through the intricate labyrinth of ethical fashion and sustainable living. Remember my initial advice in this book. Rome wasn't built in a day! Transitioning toward a sustainable lifestyle requires time, patience, and dedication. It's challenging to alter our entrenched habits, and as we've seen, societal structures don't necessarily ease the process.

Clear definitions of sustainable fashion and ethical fashion are essential because they establish a standard framework for businesses and consumers to follow, promoting transparency, accountability, and progress in the industry.

In terms of sustainable fashion, a clear definition can guide manufacturers, retailers, and consumers toward practices that minimize environmental impact. The fashion industry significantly contributes to pollution and resource depletion. We can benchmark progress toward a more environmentally friendly industry by outlining sustainability, such as using eco-friendly materials, energy-efficient production processes, or waste reduction strategies.

For ethical fashion, a clear definition is crucial in setting standards for fair labor practices, human rights, and social responsibility. This extends from ensuring fair wages and safe working conditions to advocating for diversity and inclusivity in the fashion industry. A well-defined understanding of ethical fashion ensures that all parties in the fashion supply chain are held accountable for their practices.

Furthermore, these definitions help to combat greenwashing and false claims of ethical practice, making it easier for consumers to make informed choices. In a time when consumer consciousness is rising, many are looking to align their purchases with their values. Clear, industry-wide definitions of sustainable and ethical fashion provide a roadmap for consumers to support brands genuinely committed to these principles.

Lastly, clear definitions also aid policymakers and regulators in creating and enforcing legislation to promote sustainability and ethics in the industry. With a clear understanding of these terms, developing effective laws and regulations that encourage positive change in the fashion industry becomes easier.

HOW TO SHOP FOR ETHICAL AND SUSTAINABLE FASHION

Remember the three criteria I talked about at the beginning of the book. We will apply them here to help you navigate the slow fashion world!

Criteria 1: What Is It Made From?

The path to sustainable fashion often begins with opting for clothing made from natural, organic materials. When designed in pure blends, these garments possess the potential for composting post-use. However, it's essential to acknowledge that composting clothes can present challenges, especially if you need a backyard compost system.

Many people have no idea that textiles cannot be thrown into curbside recycling bins. As a result, it's paramount to find alternative methods to keep clothing out of the garbage. I'll provide more comprehensive tips on achieving this.

Another facet of sustainable fashion lies in using synthetic materials, although this is a contentious point. While synthetics aren't the ultimate solution to our fast-fashion issues, there is a silver lining in recycling them. Keeping the existing synthetic materials in rotation can curb the production of new synthetics, thus reducing environmental harm. To put this into perspective, polyester produced for clothing emitted a staggering 282 billion kg of CO_2 in 2015 alone, nearly three times more than cotton.

Of course, the resilience of a garment is a critical consideration, and as such, durability is vital. Using safe and nontoxic dyes further safeguards the environment from the harmful effects of textile production. To validate these claims and to reassure consumers, obtaining certifications that affirm organic, recycled, or nontoxic practices is indispensable.

SUSTAINABLE FABRICS 101

Natural and organic: Natural and organic fibers are derived directly from animal or plant sources, such as cotton, linen, and wool. If certified organic, their production adheres to organic agricultural standards, which typically prohibit synthetic pesticides and fertilizers and prioritize ecological balance and conservation of biodiversity.

Organic cotton: This is cotton grown without synthetic fertilizers, pesticides, or defoliants, and its seeds are not treated with harmful chemicals. It makes up less than 0.1 percent of global cotton production. Organic cotton promotes soil health through crop rotation.

Cork: Derived from the bark of oak trees, cork fabric is an eco-friendly, vegan alternative to leather. It's recognized for its lightweight, durable, and water-resistant qualities, with a unique texture to boot.

Linen (flax fiber): Derived from the flax plant's stem, linen is a natural fiber primarily composed of cellulose. Its strength and quick moisture absorption make it ideal for warm weather, despite its susceptibility to wrinkling.

Hemp: As an ecological and fast-growing crop, hemp is simple to cultivate.

Jute: Jute, often called the "Golden Fiber" due to its golden, silky sheen, is a strong, durable, natural fiber. Its resistance to tearing, stretching, and abrasion enhances its appeal.

Bamboo fiber: Renowned for being hypoallergenic and absorbent. As a fast-growing resource, it requires fewer pesticides and fertilizers during production.

Animal fibers: Wool, cashmere, down, alpaca, silk, angora, camel, llama, mohair, and leather all fall under this category. Try to shop for upcycled animal fabrics instead of new ones or look for certifications; more on this later.

REGENERATED (SEMI-SYNTHETIC)

Semi-synthetic or regenerated fibers, primarily made from wood pulp cellulose, are produced through a process where the natural material is chemically dissolved and then reconstituted into a fibrous material for spinning. Essentially a hybrid, these fibers sit between natural and fully synthetic categories, encompassing natural fibers' comfort and breathability and synthetics' resilience and adaptability.

The conversion of these natural fibers into fabric involves a chemical "plasticizing" process. However, the sustainability of synthetic materials heavily relies on the methods and chemicals used during this conversion. Traditional viscose fabric, for instance, employs harsh chemicals like sulfuric acid in its production, which raises environmental and health concerns.

On the other hand, certain types of viscose, like lyocell, utilize less harmful solvents implemented within a closed-loop system that recycles the solvents and water. This method significantly mitigates environmental harm.

Bamboo as source material can be highly confusing. Bamboo can be chemically processed into fabric either like traditional viscose, which is less eco-friendly, or lyocell, which is relatively sustainable. This ambiguity often leads to greenwashing, making it crucial for consumers to understand the process behind the final product.

Lyocell is a semi-synthetic cellulose fiber derived from wood pulp, specifically from hardwood trees like eucalyptus and oak. It's known for its softness, strength, and high absorbency, making it ideal for clothing and bedding.

TENCEL™ Lyocell stands out as it uses REFIBRA™ technology to repurpose cotton scraps from textile waste, making it eco-friendly.

Modal is a type of cellulose fiber produced from the pulp of beech trees. It's celebrated for its softness, durability, and high absorbency, often serving as a substitute for cotton in textile manufacturing.

TENCEL™ Modal, a blend of Lyocell and Modal fibers, is derived from sustainably sourced beech trees and is lauded for its softness, durability, and moisture-wicking properties. Its eco-friendly status stems from the efficient closed-loop process used in its production.

Recycled PET (rPET): Toward the end of the 2010s, the fashion world saw a tidal shift toward adopting recycled polyester from PET plastic water bottles. This trend increased the demand for recycled PET (rPET) in clothing by an estimated 33 percent. Fashion giants began to promote recycled polyester textiles as a green alternative, but the broader environmental ramifications are more complex than the promotional narratives suggest.

One of the pressing concerns with using food-grade plastic in apparel is the intricate recycling process. Globally, recycling systems for PET differ significantly. These disparities result in difficulties like segregating food-grade from non-food-grade PET, further complicating an already challenging recycling system. While there's an abundance of PET available, the inefficiency in recycling means we're not making the most of it.

California has made strides in the textile recycling domain. The state has been proactive in implementing various textile recycling initiatives. For instance, California has had a stewardship program focused on materials like carpets for over a decade. The state collaborates with brands and local governments to devise better waste programs for hard-to-recycle products like clothing. These innovative measures make California a noteworthy player in textile recycling.

Yet, even with such advancements, there's a pressing concern about the genuine circularity of converting PET plastic bottles into garments. This transformation is marred by adding extra chemicals and materials, which can severely curtail the recyclability of the resultant clothing. Sometimes, these clothes can become entirely non-recyclable, contributing to landfill waste.

Brands highlight the environmental credentials of rPET, with some studies indicating that certain recycled polyester brands might produce up to 42 percent less pollution than virgin materials. However, for these benefits to fully materialize, there's a need to ensure that PET remains consistently recyclable throughout its lifecycle.

An unresolved query in the industry is whether rPET genuinely leads to water conservation and reduced emissions. While marketing campaigns flaunt the environmental virtues of recycled fabrics, concrete and unbiased data remain elusive. The current infrastructure, especially in the US, struggles to cope with the surging rPET demand, often leading to processing being outsourced abroad. The environmental implications of this global transportation are under-researched in terms of emissions and resource use.

Simultaneously, PET's chemical makeup raises health and environmental concerns. Chemicals like phthalates, often associated with PET, have been linked to potential endocrine-disrupting effects. Although some substances, such as terephthalates, are seen in a more benign light, others, like BPA, found even in minor quantities, can be concerning due to their hormone-disrupting properties. Additionally, PET often includes additives for protection against external elements, further adding to the list of concerns.

Findings from the Center for Environmental Health pointed to BPA's presence in several polyester-based sock brands, emphasizing that BPA concerns exist regardless of the polyester's recycled status. This discovery implies that the core issue might not be solely about recycling but also the inherent nature of polyester and its various production stages.

In conclusion, while the push to recycle PET for textiles has noble intentions, it demands a more holistic examination, ranging from its chemical components to recycling infrastructure, to ensure a sustainable impact.

ECONYL® is a regenerated nylon crafted from pre- and post-consumer waste, such as discarded fishing nets, fabric offcuts, and carpets. This recycled nylon is versatile and used across diverse sectors, including fashion, sportswear, and automotive.

LYCRA® EcoMade is a type of spandex/elastane made from pre-consumer waste. It's manufactured similarly to traditional LYCRA® but incorporates sustainable materials. Ideal for apparel demanding elasticity and recovery, it's a greener substitute for conventional spandex.

LYCRA's THERMOLITE® EcoMade T-DOWN technology turns recycled PET bottles into advanced synthetic insulation, outperforming traditional down in loft, moisture resistance, and retaining warmth when wet. The fabric is breathable and soft and quickly recovers its loftiness, positioning it as a sustainable, high-performance alternative.

UPCYCLED VS. RECYCLED CLOTHING

The main difference between upcycled and recycled clothing lies in the process used to transform the materials. Recycled clothing refers to a process where textiles are broken down into their basic components, often through mechanical or chemical means, before being spun back into yarn and used to produce new textiles. This process can degrade the quality of the materials and require significant energy use.

On the other hand, upcycling involves creatively repurposing existing garments or materials into new ones of equal or higher quality. This process doesn't break down the original materials, which often results in a more environmentally friendly process, as it typically uses less energy and maintains the quality of the original material. It also usually involves a high degree of craftsmanship and design to add value to the original items. So, while both methods aim to reduce waste and extend the life cycle of materials, recycling typically involves breaking down and remaking materials, while upcycling seeks to reuse them creatively without degradation.

INNOVATIVE ALTERNATIVES

As we endeavor to enhance sustainability within the fashion realm, we are witnessing the emergence of exceptional fabric innovations. These advancements are not just eco-conscious but ingeniously convert waste products into practical alternatives to conventional materials. Whether leveraging agricultural residues, transforming fruit waste, or exploiting nature's bounty for alternative fabrics, the fashion industry is witnessing a paradigm shift in materials sourcing and manufacturing.

AppleSkin (apple leather): This biobased material, developed from the leftover pomace and peel from the fruit juice and compote industry, provides a vegan and environmentally friendly alternative to animal leather. AppleSkin can mimic the look and feel of snakeskin, making it a popular choice for handbags and other fashion items.

Bananatex®: The world's first durable, pure fabric derived from the Abacá banana plants grown sustainably in the Philippine highlands. Known for its strength, water resistance, and breathability, Bananatex® finds use in a range of products, including bags, shoes, and clothing.

Piñatex®: Produced by Ananas Anam, Piñatex® is made from the waste leaves of the pineapple plant. Its texture and feel resemble traditional leather, making it a cruelty-free and eco-friendly alternative. Piñatex® utilizes waste from the pineapple industry and is commonly used for shoes, bags, and accessories.

QMONOS Silk and Peace Silk® represent unique advancements in silk production. QMONOS Silk uses genetic engineering to enhance silkworms, yielding silk with improved strength, elasticity, and resilience. In contrast, Peace Silk® follows an ethical approach, harvesting cocoons only after moths have naturally emerged, offering a humane alternative to traditional methods.

Cupro fabric is a regenerated cellulose material derived from cotton linter, a by-product of the cotton plant. It's often used as a silk substitute due to its smooth texture, breathability, and hypoallergenic properties.

Although made from waste product, the chemical-intensive production process raises some sustainability concerns.

Cactus leather, or Desserto, is a sustainable, plant-based alternative to traditional leather. It's made from mature leaves of the prickly pear cactus without harming the plant. This material is durable, breathable, and partially biodegradable.

Now that we've explored the diverse landscape of fabrics and their environmental implications, an essential question surfaces: As consumers, how can we confidently determine what's genuinely sustainable? How can we ensure the clothes we buy are in alignment with our values and make a positive impact on the environment? The solution rests in the domain of ethical certifications. These vital tools verify a brand's dedication to sustainable and equitable practices.

Remember, we've touched upon certifications before, specifically in the chapters concerning cleaning and laundry. Certifications certainly provide a good starting point, acting as a compass to guide us toward more ethical choices. However, like any tool, they are not perfect and have occasionally been scrutinized for effectiveness. Therefore, the onus is on us, the consumers, to peel back the layers, look beyond the surface, and conduct our due diligence to confirm the authenticity of these certifications. It's part of a larger journey toward a more sustainable fashion industry that demands conscious decision-making and critical evaluation.

COMMON CLOTHING CERTIFICATIONS

The Global Organic Textile Standard (GOTS): Certifies textile products based on their organic content and ethical production process. There are two levels: "organic" for products with 95 percent or more organic material and a second for those with 70 to 94 percent organic content. The certification covers the entire supply chain, ensuring transparency and traceability. Each GOTS label includes the license number or supplier name, allowing consumers to verify product details.

OEKO-TEX: Certifies textiles and leather for safety and sustainability. Their Standard 100 ensures every part of a product, down to threads and buttons, is tested for harmful substances. The leather standard focuses on toxic chemicals, specifically in leather. The Made in Green label, applicable to all textiles, ensures products meet chemical standards and are produced sustainably and ethically. Each Made in Green item also offers traceability for consumer transparency. While it applies to raw or finished materials, it's crucial to note that it may not cover both stages, a potential gap for greenwashing.

The Global Recycled Standard (GRS) is an optional certification that verifies the presence of recycled materials in a final product. It enforces a minimum threshold of 50 percent recycled content for consumer goods, ensuring that a brand's claim of using recycled materials goes beyond a token amount.

Better Cotton Standard (BCI): The world's largest cotton sustainability initiative. It grants certification based on adherence to seven social and economic sustainability principles.

The **Bluesign®** certification assesses products throughout their production process, from raw materials to the final product, based on environmental impact and worker and consumer safety. Products certified under Bluesign are considered environmentally friendly and free from harmful substances.

USDA Organic: This label indicates that a product was made with crops certified as organic by the USDA. However, it doesn't provide any information about the processing methods used beyond the cultivation stage.

The Forest Stewardship Council (FSC) is an international organization that certifies responsibly managed forests, ensuring they meet the highest environmental and social standards. The FSC label on a product signifies that the wood or paper-based materials used in its production come from such sustainably managed forests.

The organic nature of its materials doesn't solely determine a garment's sustainability. Its dyeing process equally matters. A t-shirt made of 100 percent organic cotton but dyed with toxic substances and vast water consumption isn't genuinely sustainable. Moreover, fabric can't be touted as biodegradable if its dyeing process involves toxins that would pollute the environment during degradation.

Fortunately, the industry is making strides in developing sustainable dyes and implementing better application methods. At a minimum, dyes should be free from azo compounds, as some can form carcinogenic aromatic amines.

Ideally, dyes should originate from plants or other natural sources, such as Colorifix, renowned for their nontoxic, non-polluting dyes that require 90 percent less water.

AirDye®: This technology revolutionizes traditional dyeing by using air instead of water. In conventional methods, water is used to apply the dye, requiring copious amounts of this precious resource and leading to potential water pollution issues. However, AirDye® technology uses a heat transfer process to shift dyes from paper onto textiles, eliminating water usage and contamination, saving energy, and reducing the number of raw materials needed. It's also capable of dyeing both sides of fabric simultaneously with different patterns, providing a unique aesthetic potential.

CO2 Dyeing: CO_2 dyeing takes advantage of a natural phenomenon where carbon dioxide behaves as a liquid when exposed to high pressure. This liquid then acts as a solvent, carrying the dye into the fabric. The process is water-free and uses reclaimed CO_2, resulting in zero waste. Additionally, it requires significantly less energy than traditional dyeing methods since the CO_2 can be heated and pressurized efficiently. Once the dyeing is complete, the CO_2 is depressurized and reverts to its gaseous state, leaving only the dyed fabric ready to be captured and reused in the next batch.

HOW TO AVOID HARMFUL CHEMICALS IN CLOTHING

Understanding & Purchasing:

- Look for certifications: Bluesign, OEKO-TEX, and GOTS indicate sustainable production and safety.
- Prioritize companies that engage in textile testing for nontoxicity.
- Opt for naturally sourced materials like linen and hemp.
- Choose clothes dyed with natural ingredients over synthetic dyes.
- Prefer undyed or neutral-colored garments to reduce chemical exposure.

Maintenance:

- Always wash new clothing in cool water before the first wear.
- Use nontoxic laundry detergents and avoid dryer sheets or choose eco-friendly alternatives.
- Install devices like PlanetCare filters to trap microfibers during washing.
- Avoid hot washes, which release more chemicals and microfibers.

Usage:

- Wear secondhand clothing as they've been washed multiple times, reducing the chemical footprint.
- Skip fast fashion and avoid "performance" fabrics (e.g., "antibacterial," "UV-protecting").
- Change immediately out of sweaty clothes post-workout.
- Transition slowly to natural and nontoxic clothing as older pieces wear out.
- If you own organic bed sheets, consider sleeping without clothing to reduce potential toxins against your skin.

Disposal:

- Donate wearable old clothes to foster reuse.
- Recycle un-wearable garments; many fabrics can be recycled, and local centers often accept them.
- Support a circular fashion industry that emphasizes reuse over waste.

CERTIFICATIONS THAT PERTAIN TO ANIMAL WELFARE

People for the Ethical Treatment of Animals (PETA) is a global organization advocating animal rights. It provides a certification that verifies products as vegan and cruelty-free, ensuring they do not contain animal-derived components and that no animal testing has been conducted in their development or production.

The Leather Working Group (LWG) certification is a sustainability standard for leather products, evaluating environmental factors like water usage, chemical handling, waste management, and wildlife protection. Four criteria assess various leather supply chain participants, leading to Gold, Silver, Bronze, or "Audited" ratings based on audit outcomes. Despite LWG's role in improving sustainability in the leather industry, criticisms surround its practices, including leniency toward hazardous chemicals use, lack of transparency, and inadequate attention to animal welfare concerns.

The Responsible Down Standard (RDS) is an independent, voluntary global standard developed to ensure the down and feathers used in goods are ethically sourced. Created through a collaboration between companies, animal welfare groups, and other industry experts, the RDS emphasizes the humane treatment of ducks and geese. One of its primary goals is to ensure traceability throughout the supply chain, certifying that the down in products comes from RDS-compliant sources. Under the standard, harmful practices such as force-feeding and live-plucking are strictly prohibited. It sets criteria for properly treating these birds, including provisions concerning their food, water, health, and outdoor access. To ensure compliance, farms undergo third-party audits.

However, despite its intentions, the RDS faces criticisms and challenges related to animal welfare. Some argue that certain harmful practices can persist or be overlooked even with the standard, especially in large-scale operations. Additionally, while third-party audits are conducted, ensuring consistent and thorough evaluations across all farms globally can be daunting. The standard's voluntary nature means that not all companies adopt it, leaving a significant portion of the down market without oversight. Some animal rights advocates argue that no down can be "cruelty-free," and I have to agree.

ZQ Merino is regarded as the highest standard of merino wool and regenerative agriculture, especially for ZQRX products. This certification is awarded exclusively to non-GMO Merino sheep farmers who adhere to stringent criteria for animal welfare, environmental sustainability, fiber quality, traceability, and social responsibility. While ZQ Merino is highly regarded, it is not flawless, and there can still be instances of animal distress. Nonetheless, it is considered the premier supply chain certification for wool. Its applicability, however, is restricted to merino sheep and does not extend to all types of wool.

The Responsible Wool Standard (RWS) is an essential certification for wool industry professionals, from farmers to merchants. It aims to enhance the welfare of sheep and preserve the grazing lands they depend on. Those awarded RWS certification are proven to adhere to stringent criteria related to animal welfare, land stewardship, and social responsibilities.

According to a study titled "Shear Destruction: Wool, Fashion and the Biodiversity Crisis," wool production has a significant environmental impact, contributing to climate change and biodiversity loss. Contrary to the perception that wool is a natural, sustainable fiber, the report argues it's a product of industrial, chemical, ecological, and genetic processes that harm the environment. It highlights that wool's climate cost is three times greater than acrylic and over five times more than conventionally grown cotton. Additionally, wool production uses 367 times more land per bale than cotton, and the chemical process of cleaning wool harms aquatic life and pollutes waterways. Despite these facts, only 28 percent of the fifty top brands using "sustainability" terms to market their wool products provided any reference to support their claims.

Allbirds, a famous shoe brand, exemplifies the problems within the wool industry. Despite its claims of sustainability and humane animal treatment, the company faces a class-action lawsuit accusing it of misleading consumers. It alleges that Allbirds understates the environmental impact of using wool and falsely represents the animal welfare practices of its suppliers. The legal action shows that despite a company's outward marketing claims, the wool production process can have significant environmental and ethical shortcomings, casting doubt on the sustainability of this natural fiber.

In light of these issues, organizations such as the Center for Biological Diversity and Collective Fashion Justice are urging the fashion industry to reduce its wool use by at least 50 percent by 2025 and focus more on innovative eco-friendly materials. They're also calling for greater transparency from companies using wool about their impact on the planet. This step will require an honest assessment of wool's substantial contribution to climate change and biodiversity loss and a commitment to shift toward more sustainable alternatives.

Both examples illustrate the complexities of living a more sustainable life. It's never black or white; there are many nuances we need to account for in this journey.

Criteria 2: Sustainable Sourcing/ Ethical Manufacturing

Ethical sourcing, also known as responsible sourcing, is an approach to acquiring goods and services that respects and upholds human rights, environmental sustainability, and social well-being. Beyond legality, it includes fair wages, safe working conditions, respect for human rights, and ecological considerations.

The importance of ethical sourcing lies in its commitment to human rights, preventing exploitative labor practices like child or forced labor or unfair wage conditions. It is a guarantor of environmental sustainability by encouraging eco-friendly materials and practices, thus reducing the harmful impacts of production processes.

Ethical sourcing often includes close attention to the supply chain, especially for businesses that rely on international suppliers. Many ethical companies prioritize smaller supply chains for greater control and transparency and strive to be knowledgeable about the countries of origin for their goods. They recognize that labor practices can vary widely by country and aim to source from places that enforce fair labor laws.

Regular audits of factories and suppliers are another critical element of ethical sourcing, providing a mechanism to ensure that suppliers consistently meet ethical standards. These audits can check for various ethical issues, from labor conditions to environmental impact, and can be crucial in identifying problems before they become significant.

Transparency is also a critical component of ethical sourcing. Companies that commit to ethical sourcing often make their practices public, allowing consumers to see exactly where their products come from and under what conditions they are made. This transparency builds trust with consumers, who increasingly value businesses that align with their ethical and social values.

Moreover, businesses that source ethically stand out in a market where consumers are increasingly conscious of social and environmental issues. This can help to build a strong brand reputation, increase consumer trust and loyalty, and ultimately contribute to a more sustainable and equitable world.

CERTIFICATIONS

Various certifications have been developed to ensure ethical sourcing that sets standards for labor practices, environmental impact, and other factors within supply chains. These certifications are instrumental in making supply chains more transparent and holding companies accountable for their practices.

One such tool is the **Fashion Revolution's Transparency Index**. This tool assesses the transparency of 250 major fashion brands annually, focusing on supply chain disclosure, environmental impact, and labor practices. Unfortunately, the 2022 Transparency Index revealed an average transparency score of a mere 24 percent, signaling that numerous brands have substantial room for improvement in their disclosure practices. Following this, several special certifications play a crucial role in enforcing ethical sourcing:

Social Accountability Standard International (SA8000): This sets the standard for fair treatment of laborers, focusing on no child labor and minimum wage payments.

B-Corporation: A business must meet high standards across eighty impact areas, both environmental and social, annually across its whole supply chain to earn this certification.

Fair Trade: Governed by various entities, it ensures reasonable prices, decent working conditions, local sustainability, and fair-trade terms for farmers and workers in developing countries.

Worldwide Responsible Accredited Production (WRAP): This certification enforces twelve labor principles, including the prohibition of forced labor, child labor, harassment, and abuse, and promotes fair work hours, non-discrimination, healthy and safe workplaces, and sustainable practices.

International Labour Organization (ILO): An agency of the UN, the ILO provides international standards to promote rights at work, encourage decent employment opportunities, enhance social protection, and strengthen dialogue on work-related issues.

Ethical Trading Initiative (ETI): An alliance of global companies, trade unions, and NGOs that promotes respect for workers' rights.

Business Social Compliance Initiative (BSCI): This industry-driven movement aims to monitor and assess workplace standards across the global supply chain.

Criteria 3: Corporate Social Responsibility

Corporate responsibility (CR), or corporate social responsibility (CSR), is an ethical framework guiding an organization's activities for long-term viability. It typically encompasses four categories: environmental responsibility, ethical/human rights responsibility, philanthropic responsibility, and economic responsibility.

One method companies use to show environmental responsibility is carbon offsetting. Carbon offsets are a form of trade; when a company buys an offset, it funds projects that reduce greenhouse gas emissions.

The benefits include:

- Promoting renewable energy.
- Encouraging businesses to reduce their environmental impact.
- Driving investment into the green economy.

However, carbon offsetting has its drawbacks. Critics argue it allows companies to buy their way out of taking immediate action to reduce their emissions. Additionally, some offset projects can have unintended negative consequences, such as displacing local communities or causing environmental damage.

Beyond carbon offsetting, businesses can employ numerous other strategies to be more environmentally responsible:

Plastic-free shipping: Businesses can use packaging materials made from recycled, biodegradable, or compostable materials.

Take-back programs: By taking back used items from customers, companies can recycle or repurpose them, reducing waste and promoting a circular economy.

Repair programs: Offering repair services prolongs the lifespan of products and decreases the need for new production.

Energy reduction: Businesses can significantly decrease their carbon footprint by implementing energy-efficient practices and using renewable energy sources.

Waste reduction: Strategies like zero-waste manufacturing and recycling programs can drastically reduce waste.

Using deadstock: Utilizing leftover or unsold stock to create new products can prevent waste and reduce the need for new materials.

HOW TO BUILD A SUSTAINABLE & ETHICAL WARDROBE

There are so many ways to create a more ethical and sustainable wardrobe! When you choose to wear, purchase, make, or start any of the practices outlined below, you are using your influence as a global citizen and a consumer in a positive way.

Make sure to check every box for every item you own! If we each do one or two of these practices every week or each time we purchase, the ripple effect will transform fashion and make it something we can love even more.

Be an outfit repeater: The most important thing you can do is to wear what you love on repeat (which allows you to shop less). Not everything needs to be worn threadbare but get all the life you can out of the items you've chosen to keep (including your fast fashion) until they no longer serve you. Up the ante by experimenting with a capsule for a few days (i.e., a limited number of items that go together). Can you use your creativity to style a different look each day, e.g., just ten items from your closet, including shoes, for the whole week? Or work toward a minimalist wardrobe with these minimalist fashion brands.

Shop less and be strategic: List new or "new to me" items you'd like in your wardrobe. Whether trendy, high fashion, or classic pieces, it doesn't matter, but ask questions about each item before you search online or in shops. For example: Will this fill a gap in my wardrobe? Do I have the budget? Is this a need or a sustainable luxury? Is it versatile? (e.g., can I wear this three ways or with three other things I already have? Does it work for layering or several seasons?) If an item is frivolous or fun, but you want the trend, see if you can find something that echoes the style at a thrift store or consignment shop. At least you are giving a second life to a garment. Often people buy trendy items on a whim, never wear them, then donate them. It's common to find things in a thrift shop that still have their tags attached! And consider renting a dress for that special occasion.

Research: Read up on your favorite brands via their corporate website. Do they say anything at all? Do they publish a corporate responsibility or sustainability report? Do they tell you what their labor or materials standards are? Alternatively, use an app like Good On You to give you a third-party assessment of brands' ethical performance and suggest better brands with similar aesthetics. Use the research of advocacy organizations like Greenpeace, Labour Behind the Label, or Fashion Revolution to help you decide who to support with your purchases.

Choose local, independent, or female-owned: Make your carbon footprint smaller by purchasing locally-made items. Keep money circulating in your community by shopping independently. Make your fashion feminist by empowering mompreneurs with your dollars.

All bodies are beautiful: Look at a brand's sizing spectrum. Inclusive sizing means that sustainable and ethical fashion is available to all. For big brands, look for at least XXS to XXL sizes, and for smaller brands that create in small batches or on-demand, look for statements that indicate "other sizes on request" or email/tag them publicly on social media to ask if they do.

Proactively support black, brown, indigenous, and people of color: According to Labour Behind the Label, "of the seventy-four million garment workers worldwide, approximately 80 percent are women" from Asian, Latin, and racialized communities in the US. Thus, ethical labor practices and organic or nontoxic processes will protect their physical health and livelihoods. However, the brands' explicit actions to dismantle systemic racism in their workplace, stores, factories, website, and social media are just as important.

Questions you can ask are: Who are their models? Who do they promote? Is there a token BIPOC here and there, or are they consistently diverse? What does their website say about their anti-racism stand and affirmative policies in place or process? Who is in leadership and staff? Finally, buy from BIPOC-owned businesses.

Mis-appropriation of cultural techniques, motifs, and styles is not an issue when you are directly supporting a BIPOC artisan or maker—be sure to ask the artisan how to wear it appropriately, and then wear their fashion and accessories proudly—and of course, share their business with all your family, friends, and followers!

Repair, tailor, or cobbler: Do you have something in your wardrobe that's been repaired, tailored, or fixed by a cobbler (shoemaker)? Get out your needle and thread to sew on that button or take them to a professional to be fitted or repaired.

The art of laundry: 70 percent of landfilled clothing has yet to be laundered correctly (shrinking, color fading, stains). Get savvy on the science of stain removal, garment care, and ironing. Conserve energy and water by washing in cold water with full loads (helps to avoid shrinking, too!). Wash clothes only when dirty, turn them inside-out, and use minimal soap. Read your care labels and wash them by hand when appropriate.

Cost-per-wear: Make a case for investing in a well-made or sustainable brand using easy math: divide the item's price by the number of wears. This is cost-per-wear (e.g., $250 dress / 20 wears = $12.50 per Wear). If you're a sale or bargain shopper, this may be a completely different way of thinking about staying on a budget, but it's an excellent mental switch to make if you want to transition your wardrobe. Buying strategically + cost-per-wear + thrifting = the slow fashionista's magic budget formula!

Vintage, consignment, thrift, or swap: Dress or accessorize with vintage items (i.e., older than twenty years), consignment, thrift, or swapped. If you have none in your wardrobe, research independent shops that are local to you or online portals for quality used items. Search for clothing swaps organized close to you, or sign up for a zero-currency exchange app, e.g., Bunz.

Rent or borrow: Do you have a special event coming up? Instead of buying something you'll only wear once, borrow from a friend, a peer-to-peer service, a fashion library, or a rental company.

Repurpose/reuse: Make or purchase something repurposed, e.g., jeans that are now shorts, a leather couch that is now a bag, bicycle inner tubes that are now earrings, and bullets that are now rings. Reuse of textile ends and "deadstock" is now becoming more common. With more organizations coordinating waste, such as The Queen of Raw, it's easier for brands to produce collections in small batches based on how much textile they can obtain. These collections often sell out, which also reduces waste.

Nothing to landfill: Any unworn and unwanted items can be donated next time you do a closet edit. Research a local organization requesting specific clothing items for donation (e.g., women's office wear for women coming out of the shelter system; winter jackets at a nearby school), or call a local charity for pickup. Nothing should go in the garbage, not even holey socks, underwear, or bras. Everything can be reused, repurposed, or recycled, so donate it all. Just be sure it is clean and dry.

> Decluttering can be overwhelming due to the significant time investment it often requires. I've devised a simple yet effective strategy in response to this common concern. After attempting to declutter my home and wardrobe, I introduced a small basket at the bottom of my closet. Over time, I would put clothes I no longer wore or wanted into this basket. After several months, the basket was full, signifying it was time to donate these items. This method eliminated the need for large, time-consuming closet purges. This strategy can be equally effective with children's clothes. When an item no longer fits, it goes directly into the donation basket instead of returning it to the wardrobe. Over time, this practice makes it easy to identify and remove outgrown clothes. You can apply this system in any room, not just bedrooms and closets. Whenever you come across something you no longer need or want, place it in a dedicated box. Once the box is full, you'll know it's time to donate, recycle, or responsibly dispose of these items. This method simplifies what can feel like a monumental task and encourages mindful consumption.

In the evolving world of ethical fashion, two prominent regions are leading the way: New York and France. Both champion sustainable and ethical practices that resonate deeply with an informed consumer base, signaling a tectonic shift in the fashion industry's ethos.

New York is positioning itself at the forefront with the **Fashion Sustainability and Social Accountability Act**. Targeting apparel and footwear companies with a global revenue exceeding $100 million that operate within its jurisdiction, this act is an attempt to leverage New York's vast market size, urging brands worldwide to adopt more sustainable practices if they wish to engage with their consumer base. The legislation places heavy emphasis on supply chain transparency. Companies are now mandated to disclose the entirety of their supply chains, from raw materials to finished products. But it doesn't end with transparency. The act's Mandatory Due Diligence Framework compels companies to reduce their negative environmental and social impacts actively. This includes adhering to the Paris Agreement, collaborating with suppliers to manage chemical usage, and implementing concrete steps to uplift the lives of garment workers, especially in terms of fair remuneration. These guidelines are enforced by the New York Department of State and the attorney general. Non-compliant companies face hefty fines, with the resulting funds channeled to benefit the communities the fashion industry has historically exploited.

Parallelly, France is sustainably making strides. The French government recently announced a scheme to subsidize repairs of clothing and shoes aimed at reducing waste and curbing pollution from the textile industry. This innovative initiative, introduced by Secretary of State for Ecology Bérangère Couillard, offers varying discounts for clothing and shoe repairs. For instance, resoling a shoe can fetch a €25 rebate, promoting product longevity and discouraging wastefulness. As Couillard expressed, the overarching goal is to cultivate a circular economy, enhancing the lifespan of products and endorsing the principle of product reincarnation. The alarming statistic that the textile industry could account for a quarter of global greenhouse gas emissions by 2050 further underscores the urgency of such initiatives. To operationalize this, the Ministry of Ecology has partnered with a private organization called Refashion. Notably, businesses can join this initiative for free, with Refashion collecting a minimal "eco-contribution" on sales to fund the subsidy, ensuring the consumer directly benefits.

Canada's new law on forced and child labor in supply chains, Bill S-211, is criticized for being ineffective. Critics, including the Canadian Network on Corporate Accountability (CNCA), argue that the law only mandates companies to report on their actions against forced or child labor but doesn't require them to take concrete steps if such practices are found. Human rights groups and labor activists believe this approach is insufficient, as it does not hold companies accountable or assist victims. They stress the need for a more robust law that includes mandatory human rights and environmental due diligence, allowing victims access to remedies through Canadian courts. Furthermore, the law is seen as failing to help Canadian consumers avoid products made with forced or child labor, nor does it support Canadian companies actively combating human rights abuses.

Historically associated with aesthetics, fashion is now intertwined with sustainability and ethical consciousness. As Stella McCartney emphasized, there's a dire need for the fashion industry to commit to action that paves the way for a sustainable future. With measures such as those introduced by New York and France, the global narrative around fashion is slowly but surely shifting toward a more compassionate, sustainable, and ethical paradigm.

CHAPTER 6

Skin Deep: The Real Impact of Sustainable Beauty

Natural! Clean! Green! Nontoxic! Stroll down any cosmetic section, and you're bombarded with assurances of products that claim to be everything your skin needs. They promise to be the gentle, caring companion your skin yearns for. However, these buzzwords hold little to no concrete meaning in the larger scheme of things. Much like the term "natural" that came before them, "clean," "green," and "nontoxic" are not regulated by any governing bodies. This lack of regulation allows brands to use these terms indiscriminately, creating confusion about what you're investing in when choosing a beautifully packaged "clean beauty" product that promises exceptional skincare.

However, that doesn't imply that the clean movement is all just smoke and mirrors. Despite a few questionable players exploiting these terms, most consumers and retailers have become savvy enough to spot such gimmicks. We've had our fair share of learning through the greenwashing fiasco of the 2010s. The real puzzle with "clean" lies in its definition, which varies widely based on individual perspectives. Does "clean" only include natural ingredients, or do synthetics with a proven safety record make the cut? Are only sustainably sourced ingredients deemed acceptable? And how does plastic packaging fit into the "clean" narrative? And ultimately, who is the authority on deciding what's safe?

Regrettably, these questions remain unanswered as regulatory bodies have yet to step in, leaving us to navigate this maze ourselves. Consumers now need to channel their inner detective skills to scrutinize the origins of preservatives, ensuring they meet individual safety and environmental standards. Quite a task, right?

The silver lining is that many brands and retailers, acknowledging consumer frustrations, have taken it upon themselves to clarify their interpretation of "clean." This increased transparency can assist you in determining whether a product aligns with your clean beauty standards.

In today's market, companies have an alarming freedom to make bold claims without substantiating them or even following through. With the surge of the clean beauty trend, it seems like every brand is vying to slap labels such as "safe," "natural," and similar adjectives onto their products. However, this isn't an issue confined to clean beauty alone. Even terms like "clinical" and "science-backed" aren't strictly regulated.

Many brands are likely to use "clean" with the best intentions. Still, given its vague definition, it's left to each brand's discretion to interpret what they think "clean" genuinely signifies. This leads to specific ingredients or preservatives receiving undue criticism due to marketing tactics, despite being perfectly safe from a scientific standpoint.

Regardless of the varied perspectives on the nuances of clean beauty, a central theme seems to resonate with everyone. A beauty product is deemed "clean" if every ingredient and sub-ingredient in it is proven safe for both people and the planet. This overarching idea resonates with many in the industry, although interpretations of "safe for the planet" can range widely. It might refer to practices such as sustainable harvesting, limiting virgin plastics in packaging, or even reusable packaging systems. Before we dive into my criteria to help you shop for sustainable and ethical skincare, we need to understand the many issues facing the industry today.

1: PLASTIC PACKAGING PREDICAMENT

It's a huge problem and plays a significant role in the sustainability conversation surrounding the beauty industry. Believe it or not, 70 percent of the beauty industry's waste is attributed to packaging.

To put it into perspective, each year, the global cosmetics industry produces an astonishing 120 billion packaging units. The real shocker? Most of this packaging doesn't end up being recycled. If we don't seriously address this issue, scientists warn that by 2050, we could face a staggering 12,000 metric tons of plastic waste, either filling our landfills to the brim or devastating our natural environment.

Considering this alarming reality, it's no surprise that we need to scrutinize what's inside our makeup bags more carefully. But the issue of packaging extends far beyond our cosmetic collections. The implications ripple out to the broader ecosystem, affecting biodiversity, contributing to climate change, and posing a significant threat to our waterways and marine life.

Moreover, we need to examine the lifecycle of these packaging materials. From sourcing the raw materials to the energy-intensive production process and finally to the disposal, each stage poses significant environmental challenges. It's crucial to consider not just the end of life of these packaging materials but the entirety of their lifecycle.

Then there's the question of convenience versus necessity. Single-use products, excessive wrapping, and non-refillable containers are all driven by a desire for convenience rather than need.

2: INGREDIENT TRACING + TRANSPARENCY

The cornerstone of sustainability in beauty is an in-depth understanding of ingredient origin. It is fundamental for any beauty brand genuinely committed to sustainability to have precise knowledge about the origins of its ingredients. Without this insight, it's impossible to discern how ingredients were grown, harvested, processed, treated, or shipped. This knowledge gap could mean the difference between crafting a beauty product that embodies sustainability and one that inadvertently causes harm.

A critical ingredient presenting substantial sustainability challenges is essential oils. Essential oils typically require a significant land footprint for a minimal yield. The recent surge in essential oil network marketing has caused consumption to skyrocket, leading to overuse. Many of these ingredients are also slow-growing, adding to the sustainability dilemma. How some of these ingredients are harvested can also amplify the issue. For instance, the extraction of frankincense involves tapping Boswellia trees, which can be extremely damaging. To meet the growing demand, the extraction process is being accelerated, leading to the overexploitation of these trees.

Almond oil is another ingredient gaining traction in beauty products, but it brings unique challenges. Almond cultivation relies heavily on bees for pollination. An overwhelming majority of the world's almonds hail from California, where native pollinators can't meet the demands of the almond-growing industry. As a result, beekeepers have to ship their bees in to assist in the pollination process. A survey of commercial beekeepers revealed that a staggering fifty billion bees were wiped out over a few months during the winter of 2018–2019. This mass die-off was linked to many factors, including pesticide exposure, diseases from parasites, habitat loss, and reliance on industrial agricultural methods.

Argan oil, extracted from the fruit of the Argania Spinosa tree, is widely used in hair and skin care products due to its hydrating and shine-enhancing qualities. However, the argan tree is under threat because of livestock interference that hinders its growth and the demand for its wood for construction purposes. Notably, it takes over half a century for an argan tree to bear fruit fit for oil extraction. While numerous beauty companies obtain argan oil from sustainable cooperatives that support local communities, there are concerns about some firms potentially choosing less expensive and less sustainable alternatives.

Bakuchiol, a natural alternative to retinol, has gained popularity in the skincare industry. It is derived from the Psoralea Corylifolia plant and is praised for its ability to smoothen, clarify, and brighten skin. Although it is not currently listed as vulnerable or endangered on the International Union for Conservation of Nature (IUCN) Red List or the Convention on International Trade in Endangered Species (CITES), there are reports highlighting the potential risk of endangerment due to unchecked collections. Despite being a key component of many skincare items, the sustainability of its origins remains questionable.

3: HUMAN RIGHTS

The process of procuring raw materials frequently used in beauty and personal care items is increasingly being examined due to the high risk of child labor in mining and agriculture, especially in the lower levels of the supply chain. The mining of mica—a mineral primarily found in India, Madagascar, Brazil, and China and used to provide shimmer and color in products like eyeshadows and blushes—is a striking example in this industry. Mica mining often occurs in hazardous and unregulated environments, typically in areas with lax labor regulations. Around a quarter of the global mica supply comes from Jharkhand and Bihar, India, where over 22,000 children work in mines, violating their rights to safety, health, and education. Other raw materials such as vanilla, cocoa, copper, and silk, which feature in the manufacturing of foundations and creams, have also been linked to child labor in some of their primary producing countries (for instance, child labor has been reported in cocoa production and supply chains in Ghana and Côte d'Ivoire). Companies should strive to source their raw materials responsibly by collaborating with suppliers to enhance supply chain transparency and traceability to tackle and eliminate the root causes of child labor.

Additionally, it's worth noting that nearly 30 percent of ingredients in beauty and personal care items come from either mined or agricultural commodities like copper, mica, and palm oil. These commodities are often classified as "high risk" due to well-documented labor rights abuses in poorly regulated sourcing operations. These sectors also depend heavily on low-skilled labor, often recruited through agencies, resulting in potential issues such as passport confiscation, excessive recruitment fees, debt bondage, and withheld contracts and wages. Workers involved in the production of agricultural raw materials like palm oil, carnauba wax, or candelilla wax, especially in countries like Indonesia, Brazil, and Mexico, are often subject to adverse conditions, such as forced labor, low pay, extended work hours, sexual harassment, health and safety risks, discriminatory practices, and limited or no access to complaint mechanisms.

Mica is a naturally occurring mineral that adds a sparkling effect to many cosmetic products. In contrast, synthetic mica, or synthetic fluor phlogopite, is produced in a laboratory setting.

The key difference between natural and synthetic mica lies in their sourcing and purity. Natural mica is mined from the earth, and this process can often introduce contaminants such as heavy metals like lead and mercury. In contrast, synthetic mica is produced in controlled environments, ensuring it's free from contaminants. This not only makes synthetic mica purer but also safer for consumer use.

Furthermore, synthetic mica manufacturing has a reduced environmental impact compared to natural mica mining. Natural mica extraction often involves deforestation and waterway pollution, whereas synthetic mica is created from natural materials like plant cellulose, ensuring no harmful microplastics are released into the environment.

Despite the benefits of synthetic mica, the cosmetics industry still predominantly uses natural mica due to cost factors. Additionally, many cosmetic companies lack supply chain transparency, making it challenging for consumers to know whether the mica in their products is ethically sourced.

4: THE AIR AND WATER POLLUTION CRISIS

When we use our beauty products, we may not realize that we could be releasing harmful substances into our environment. Many chemical ingredients, including BHT, Sodium Laureth Sulfate, and BHA, present in everyday skincare and beauty items, have profoundly disrupted aquatic life. These chemicals, even in minute concentrations, can drastically alter the biochemistry of marine organisms, reducing plankton populations and sometimes even resulting in fish fatalities.

Perhaps more concerning is that these chemicals persist in our water supply, even after extensive sewage treatment. This persistence points to the insidious long-term impacts on our environment, raising serious questions about the safety and sustainability of these common ingredients.

Simultaneously, the beauty industry is also a significant player in air pollution. A key contributor is Volatile Organic Compounds (VOCs), found in many beauty products such as perfumes, deodorants, and hairsprays. Recent studies have unveiled that VOC emissions from household and beauty products account for a sizeable proportion of total VOC emissions in urban areas, contributing significantly to carbon dioxide emissions and exacerbating the challenges of climate change.

We must also consider the carbon footprint resulting from the fossil fuel-derived materials used extensively in beauty products and their packaging. The global cosmetics industry's reliance on these materials contributes significantly to carbon pollution. Unsustainable extraction and processing methods can lead to considerable carbon emissions, even regarding natural ingredients.

5: UNMASKING DEFORESTATION

The beauty industry and food production rely heavily on plant-based raw materials, a significant portion of which comes from plantations established on lands cleared of natural forests. In addition to this, the packaging for beauty products—such as cardboard boxes, paper wraps, and packing tissues, all derived from trees—contributes significantly to deforestation. Annually, producing these packaging materials results in the loss of about eighteen million acres of forest.

Palm oil is a primary contributor to this deforestation. A recent study revealed that in Southeast Asia, nearly 50 percent of oil palm plantations have replaced forests since 1989. These figures rise to 54 percent and 40 percent in specific countries such as Indonesia and Malaysia.

The environmental implications of this rapid deforestation are severe. Trees play a vital role in carbon sequestration, absorbing CO_2 from the atmosphere. When they're cut down, this CO_2 is released back into the atmosphere, thus exacerbating global warming. Moreover, this widespread deforestation drastically impacts biodiversity. Native plants and wildlife lose their natural habitats, resulting in a sharp decline in their populations. A startling example is the 68 percent decrease in wildlife populations since 1970, a trend that disturbingly continues.

Palm oil production is particularly detrimental to forests of high conservation value, with plantations leading to increased greenhouse gas emissions, soil erosion, and water pollution. This habitat destruction puts almost two hundred species at risk.

The industry also faces criticism for social issues, including rampant labor rights violations. Reports of child labor and gender discrimination are frequent. Children are often compelled to work without safety equipment and are exposed to harmful chemicals. At the same time, women, often employed as casual laborers, are denied benefits that accompany permanent employment, such as health insurance and pensions.

Organizations like the Roundtable for Sustainable Palm Oil (RSPO) attempt to regulate the industry by setting sustainable palm oil production standards. However, their criteria have been criticized as vague and open to interpretation. Furthermore, critics argue that the predominance of industry representatives among the RSPO's members could lead to potential biases and inadequate enforcement of standards. Accusations of collusion between third-party auditors and palm plantations to hide violations also exist.

Although palm oil production offers employment opportunities in lower-income countries, the lack of stringent regulations and enforcement leads to accusations of greenwashing in the current model of sustainable palm oil. While alternatives like sunflower oil, grapeseed, and coconut oil exist, they are less likely to be adopted due to the higher yield of palm oil. Some companies offer palm oil-free products, allowing consumers to choose not to contribute to the issues associated with palm oil production.

Rosewood essential oil, sourced from the Aniba Rosaeodora tree, is a popular ingredient in aromatherapy and skincare, but it's currently classified as endangered on the IUCN Red List. The use of rosewood has recently become problematic. Even though its application in beauty might be relatively small compared to other industries, its environmental impact is substantial and should not be dismissed. The extraction of rosewood oil requires the entire tree to be felled. Given that rosewood is a slow-growing tree, this extraction method poses significant sustainability challenges and contributes to deforestation, a leading cause of climate change.

6: ANIMAL RIGHTS

In an increasingly conscious world, animal testing issues in the cosmetics industry have gained attention and fostered significant change. Despite not being a requirement in many countries, animal testing has been a controversial feature of the beauty industry. Countries like the UK, EU, India, Israel, Australia, New Zealand, Turkey, Switzerland, Norway, and most recently, Canada have enacted laws banning or limiting this practice.

Canada's amendment to the Food and Drug Act, included in this year's budget, reflects a decision to ban cosmetic animal testing, putting it in line with forty-three other nations.

However, the issue is not entirely resolved. The term "cruelty-free" often applied to beauty products can be ambiguous. It may refer to the final product not being tested on animals, but this does not guarantee that individual ingredients sourced from suppliers are also free from animal testing. Sadly, most skincare ingredients have undergone animal testing at some point. It's also worth noting that "cruelty-free" does not necessarily equate to "vegan." Animal-derived ingredients can still be present in cosmetics labeled as "cruelty-free." Similarly, "vegan" products, while free from animal-derived ingredients, may still undergo animal testing. These complex realities underscore the need for consumers to remain vigilant and informed.

Look out for these ingredients on the label:

1. **Keratin:** Sourced from the hair and horns of various animals, keratin is found in shampoos and conditioners to strengthen hair and nails.
2. **Tallow:** This animal fat is used in many nail polishes, soaps, eye makeup, and foundations. Also known as oleic acid, oleyl stearate, and oleyl oleate.
3. **Beeswax (Cera Alba):** Commonly used to produce lip balms, soaps, and moisturizers. Beeswax enhances the skin's absorption of moisture.
4. **Guanine:** Used to create shimmering effects in eyeshadows.
5. **Lanolin:** Derived from sheep's wool, it's used in lip balms, lipsticks, and lip glosses.
6. **Carmine:** Provides the vibrant red color in lipsticks, blushes, and nail polishes. It's derived from crushed cochineals and can be identified as natural red 4, E120, and C.I 75470.
7. **Shellac:** Sourced from lac bugs, shellac adds shine and strength to nail polishes.

7: MICROBEADS = MAJOR TROUBLE

Microbeads are teeny-tiny but pack a punch regarding environmental damage.

These microscopic particles are found in various personal care products—from facial scrubs, toothpaste, and body wash to other unsuspecting products like lipstick, eyeliner, sunscreen, deodorant, and even nail polish. Their abrasiveness also lands them a role in various household cleaning items.

What's the big deal with these microbeads? Well, they're plastic. And when we rinse them off, they can slip through water treatment facilities and end up in our rivers, lakes, and oceans, causing harm to aquatic life and polluting our waterways.

Acknowledging the problem, several countries have banned these harmful particles. The US, for instance, passed the Microbead-Free Waters Act in 2015, which phased out microbeads in rinse-off cosmetics by July 2017. Other countries, including Canada, France, New Zealand, Sweden, Taiwan, and the UK, have banned microbeads from rinse-off cosmetics.

Canada has made significant strides. In July 2019, the country prohibited selling and manufacturing toiletries containing microbeads. Then, in May 2021, it went a step further, adding plastic-manufactured items, including microbeads, to its list of toxic substances under the Canadian Environment Protection Act.

Exploring ethical and sustainable beauty is imperative after delving into the various issues plaguing the beauty industry, but before we move forward, we need better definitions of each. **Ethical beauty** refers to products and practices prioritizing the well-being of people, animals, and the environment. It ensures fair wages, safe working conditions, and human rights protection throughout the supply chain. Additionally, ethical beauty entails using cruelty-free ingredients, avoiding harmful chemicals, and promoting transparency in sourcing and manufacturing processes. **Sustainable beauty** goes beyond ethical considerations by emphasizing long-term environmental sustainability. It entails using renewable resources, reducing carbon footprint, minimizing waste generation, and adopting eco-friendly packaging. Ultimately, ethical and sustainable beauty seeks to redefine the beauty industry by promoting conscious consumer choices and fostering positive social and environmental impact.

HOW TO SHOP FOR ETHICAL AND SUSTAINABLE SKINCARE PRODUCTS

By now, you know how much I love criteria. They offer a framework for navigating the complexities of a more sustained life. Regarding beauty and skincare, I follow the following:

1. Better Ingredients

The debate surrounding using "chemicals" in beauty products is complicated and often misunderstood. When most people refer to "chemicals," they're usually talking about harmful compounds or synthetic ingredients. However, scientifically, everything around us is made of chemicals, including natural elements and water.

In beauty and skincare, certain brands have capitalized on the fear of "chemicals," promoting their products as "chemical-free." It's crucial to note that this term isn't entirely accurate. Every substance is technically a chemical. Claiming a product is "free of chemicals" is often more about marketing than reality and can, unfortunately, contribute to misinformation.

The distinction must be made not between "chemicals" and "non-chemicals" but between harmful and safe ingredients. These can be either synthetic or natural. The misconception that all natural means safe is misleading. Numerous naturally occurring substances, such as lead, mold, and poison ivy, can be harmful. Similarly, not all synthetic ingredients are harmful. Our bodies constantly synthesize ingredients from the nutrients we consume.

It's not the source of an ingredient (natural or synthetic) that determines its safety but its nature, quantity, and how it's used. Reading and understanding ingredient lists should be your guide to the safety of beauty products. The key here is to avoid known toxic compounds, whether naturally derived or synthesized.

The sustainability debate has drawn a line between natural and synthetic ingredients, each with merits and drawbacks, particularly in the skincare sector that dominates the cosmetics market.

While natural brands argue that their ingredients, drawn directly from the earth, are inherently sustainable, some say that these natural resources can be depleted over time, calling into question their sustainability. Brands with a more synthetic approach maintain that their products, derived in laboratories, avoid draining natural resources.

As consumer demand for sustainable goods rises, brands are adapting. Those who don't incorporate sustainability into their ethos may struggle to stay relevant in the coming years. The importance of this shift is further highlighted by the ongoing climate crisis, which makes sustainability a necessity rather than a trendy buzzword.

A closer look at the natural versus synthetic debate reveals complexities. There are concerns about the environmental impact of harvesting large quantities of natural resources to create beauty products. On the other hand, artificial ingredients, like a lab-created vanilla scent, may be more sustainable than growing vanilla, but their overall environmental friendliness requires deeper analysis.

The economic implications of these ingredients are also significant. Many communities depend on producing natural ingredients like shea butter, seaweed, coconut oil, and moringa oil. Supporting these plants' sustainable trade can help protect biodiversity and drive environmental conservation.

In the bid to strike a balance, biotechnology offers promising solutions. Brands like **One Ocean Beauty** and **Biossance** utilize bio-fermentation and genetic engineering techniques to reproduce natural molecules in a lab, thus creating active ingredients without negatively impacting the environment.

In the end, assessing the sustainability of an ingredient isn't simply about whether it's natural or synthetic but involves a comprehensive understanding of the ingredient's life cycle and environmental, social, and economic impacts. Transparency and consumer education are vital in making informed choices in this evolving market. Companies like **Novi Connect** play a crucial role in the clean beauty movement by providing platforms that assess the sustainability of raw materials. These platforms use technology to evaluate each ingredient based on factors such as its environmental impact, safety, and ethical sourcing.

Ultimately, consumers are encouraged to align with brands that prioritize sustainability and are transparent about their practices. Such initiatives will help increase the volume of sustainable products available, reassuring consumers that the ingredients in their beauty products meet the sustainability standards they care about.

Specific ingredients found in conventional beauty products should be avoided. Many are linked to serious health risks for people and the planet and are petroleum-based, and I don't know about you, but I don't want petrol anywhere near my skin.

THE UNCLEAN EIGHTEEN—INGREDIENTS TO AVOID IN SKINCARE

1. **Sodium Lauryl Sulfate (SLS), Sodium Laureth Sulfate (SLES), Ammonium Laureth Sulfate (ALS):** Surfactants in washes and toothpaste. Can cause irritation, and potential contaminants may be carcinogenic.

2. **Parabens:** Preservatives in hygiene and beauty products. Can disrupt hormones, cause skin irritation and is linked to cancer.

3. **Phthalates:** Additives in various products. Potentially linked to DNA damage, organ damage, and neurodevelopmental effects.

4. **Synthetic fragrance/parfum:** In most products to add scent. Can emit pollutants, linked to several health issues.

5. **Ethanolamine compounds (MEA, DEA, TEA):** Surfactants and pH balancers in hygiene products. Prolonged exposure can affect health.

6. **Synthetic colors (FD&C or D&C):** Enhance product appearance. May contain heavy metals and pose health risks.

7. **Triclosan and Triclocarban (TSC and TCC):** Antibacterial agents. Can lead to antimicrobial resistance, hormonal imbalance, and environmental issues.

8. **Polyethylene Glycols (PEGs):** Penetration enhancers in hygiene products. Potentially contaminated with carcinogens and linked to organ toxicity.
9. **Paraffin/petrolatum:** Emollients in beauty products. Can cause skin irritation, collagen damage, hormonal imbalance, etc.
10. **Coal tar:** Skin treatment and colorant. Classified as carcinogen with potential links to skin cancer and other conditions.
11. **Siloxane:** Softeners in various products. Can disrupt hormones, harm the immune system, and pose environmental harm.
12. **EthyleneDiamineTetraacetic Acid (EDTA):** Enhances foaming and extends shelf life. Can cause cell and gene toxicity and skin irritation.
13. **Chemical UV filters (Octinoxate and Oxybenzone):** Sun radiation blockers. May disrupt hormones, pose reproductive risks, and harm the environment.
14. **Formaldehyde and formaldehyde-releasing preservatives (FRPs):** Preservatives. Carcinogenic and can cause dermatitis.
15. **Butylated Hydroxyanisole (BHA) and Butylated Hydroxytoluene (BHT):** Synthetic antioxidants and preservatives. Linked to organ toxicity, skin irritation, cancer, and endocrine disruption.
16. **Methylisothiazolinone and methylchloroisothiazolinone:** Preservatives. Can be corrosive, allergenic, toxic to organs, and harmful to aquatic life.
17. **Benzalkonium chloride:** Surfactant, stabilizer, preservative, antibacterial. Toxic and allergenic to skin.
18. **Talc:** Absorbs moisture and prevents caking in powder products. Potential links to several cancers and can contain asbestos.

CERTIFICATIONS

Turning to third-party certifications provides a trustworthy compass to navigate the complex landscape of clean beauty. These certifications assure consumers that the products they choose are produced under stringent standards for organic content, environmentally friendly manufacturing, ethical sourcing, transparency, and the absence of genetically modified organisms, thereby supporting health and sustainability.

USDA Organic: A certification from the United States Department of Agriculture indicating a product is made of at least 95 percent organic ingredients and adheres to strict farming and production practices regulations.

Soil Association Organic: This UK-based certification assures a product meets high standards for organic content, ethical sourcing, and environmentally friendly manufacturing.

Ecocert: A French certification body that validates a product's organic content, sustainability, and environmentally friendly production methods.

COSMOS: A certification that combines organic and natural standards across Europe, emphasizing environmental management, green chemistry, and consumer transparency.

GMO-free: A general term for products free from genetically modified organisms (GMOs).

Non-GMO Project Verified: A North American certification indicating a product and its ingredients have been thoroughly tested and found to be free of genetically modified organisms.

EWG Verified: A seal from the Environmental Working Group that signifies a product meets their strict criteria for transparency and health, avoiding ingredients EWG finds concerning.

2. Sustainable Sourcing/ Ethical Manufacturing

In issue 2, I mentioned the importance of ingredient traceability. This is the next phase of the green beauty revolution. A growing trend of shoppers trying to discern the difference between potentially harmful, mass-produced ingredients like palm oil, coconut oil, and low-cost fractionated oils versus wholesome, nutrient-dense extractions and oils. The crux of the matter is that the intricate blend of natural and synthetic materials in beauty products brings about considerable challenges regarding sustainability and traceability. Given that natural ingredients are obtained from diverse sources, including plants, animals, and the earth, multiple parties like farmers, wild harvesters, miners, ingredient brokers, cooperatives, and import-export entities are involved in the supply chain. This vast network makes it difficult to identify the precise origins of ingredients. For more prominent brands that work with numerous suppliers and hundreds of smaller suppliers, tracking the source of ingredients becomes even more daunting. Moreover, traceability plays a critical role in ensuring product safety. When the exact contents of a beauty product remain unknown, risks such as contamination, unanticipated allergic reactions due to undeclared ingredients, and ethical concerns like animal testing might arise within the supply chain. Therefore, traceability increases certainty about the product's safety for consumers, manufacturers, and animals alike.

Organizations like **Sourcemap** play a vital role in helping brands enhance their supply chain transparency by tracking certified raw materials and ensuring equitable labor conditions and wages. Sourcemap provides a mechanism for a brand to map and verify the sourcing and production of each ingredient in all products in a consumer-friendly manner.

This service proves particularly useful for beauty brands that use mica, as Sourcemap's Responsible Mica Platform pinpoints every mine and processing facility in the mica supply chain. They conduct third-party audits of suppliers for each mineral shipment. They offer a similar mapping process for palm oil, spanning hundreds of thousands of plantations and an intricate processing and export supply chain. By sharing satellite images of the plantations of origin with consumers, this service helps palm oil product users better understand the significance of sourcing sustainably without contributing to deforestation.

Moreover, several other traceability solutions exist, such as certifications and deep or vertical integration. Certifications often communicate a brand's environmental and social commitments and may shed light on the origin of specific ingredients. For instance, the

Unique Manuka Factor (UMF) certification assures that the manuka honey used in a product is genuine. On the other hand, the Roundtable on Responsible Palm Oil (RSPO) employs blockchain technology to trace the supply chain of palm oil and its derivatives.

Deep integration is an appealing option for many brands, as it minimizes the number of participants in a supply chain. In this model, brands either own and operate farms supplying their raw materials, or collaborate with a sole supplier who farms and processes the crops into natural compounds, which are then directly utilized by the brand to manufacture their products.

While improving traceability and transparency, this model fosters better working conditions for farmers and promotes increased sustainability as suppliers care for the local community and the environment where ingredients are grown and processed. However, consumers need to trust that brands are not merely fabricating marketing ploys but also maintaining direct supplier relationships. Consumers can uncover discrepancies in a brand's narrative by asking a few questions, as genuinely sustainable operators are eager to share information about their relationships with farmers and communities.

The ISO 22005 standard outlines guidelines for enhancing traceability in the food and feed chain, aiming to improve transparency and sustainability. Although initially designed for the food industry, its principles can also be used in the beauty industry. Applying this standard would allow beauty brands to track and document the journey of each ingredient from source to final product. This can boost consumer trust by providing clear information about ingredients' origins and verifying sustainable and ethical sourcing claims. Essentially, ISO 22005 could elevate supply chain management and encourage better practices in the beauty industry.

We're still waiting for a universally accepted approach for a more thorough industry-wide standard for beauty products. In the interim, consumers might prefer products that carry the **MadeSafe**™ certification. While this nonprofit organization doesn't tackle traceability directly, it does use a team of scientists to thoroughly analyze a product's molecular composition, ensuring the safety and suitability of its ingredients for the intended use. This serves as an initial assurance of traceability for consumers. Nevertheless, brands must provide more extensive details about these verified ingredients' origins.

CERTIFICATIONS

Fairtrade: Fairtrade International certifies fair trade for underprivileged producers in developing countries, ensuring minimum pay and a Fairtrade Premium for local investment. Its mark vouches for this.

Fair Trade: Fair Trade USA, distinct from Fairtrade International but similarly missioned, guarantees safe, humane work environments, prohibits child/forced labor, ensures fair wages, and maintains a community development fund. Its Fair Trade Certified™ label is proof.

3. Cruelty-Free

Choosing cruelty-free skincare is an essential aspect of ethical shopping. Cruelty-free products aren't tested on animals, promoting humane practices in skincare development. Technological advancements now provide alternatives to animal testing, making it unnecessary and often less reliable due to different physiological responses between humans and animals. Cruelty-free brands like **Meow Meow Tweet** and **Axilolgoy** illustrate that high-quality skincare can be produced without animal testing. Consumers can encourage other brands to adopt similar ethical practices by choosing these products. Moreover, cruelty-free often aligns with other green practices like sustainable sourcing and eco-friendly packaging. In essence, opting for cruelty-free skincare supports both animal welfare and environmental sustainability.

CERTIFICATIONS

Leaping Bunny: Managed by the Coalition for Consumer Information on Cosmetics (CCIC), the Leaping Bunny Program certifies companies and brands that refrain from animal testing at all stages of product development.

PETA (People for the Ethical Treatment of Animals): PETA's "Beauty Without Bunnies" program offers cruelty-free and vegan certification for brands that pledge to abstain from animal testing and avoid all animal-derived ingredients. Also found on vegan products.

Choose Cruelty-Free: An Australian-based organization, Choose Cruelty-Free provides certification for brands that neither test their products on animals nor source ingredients from animal testing suppliers.

4. Vegan

Vegan beauty products, devoid of animal-derived ingredients, play an essential role in supporting humane and eco-conscious practices. However, choosing brands that offer vegan alternatives using ethically sourced ingredients is necessary. Despite their plant-based nature, ingredients like palm oil can negatively impact the environment and disrupt animal habitats. Additionally, "vegan" products may contain nonorganic or synthetic elements, indicating that "vegan" doesn't necessarily mean "natural" or "organic." Not all vegan products ensure cruelty-free standards, emphasizing the need to seek both labels for truly animal-friendly choices.

Transitioning to a related issue concerning biodiversity, recent research has spotlighted the impact of urban beekeeping on wild bee populations. This study, conducted by Concordia University, indicates that the rise in urban beekeeping activities is distressing the local wild bee species. Due to their superior foraging abilities, the surging honeybee populations are out-competing wild bees, leading to their decline. This situation has been particularly acute in Montreal, where urban hives have grown twelve-fold in less than a decade, negatively impacting wild bee populations.

The research team suggests a precautionary measure of maintaining three hives per square kilometer, a threshold currently being exceeded in Montreal. The significant increase in urban beekeeping might endanger local biodiversity and necessitate a more balanced approach. For instance, planting pollinator gardens is recommended as a more effective method to boost urban biodiversity instead of adding more hives.

While urban beekeeping has its merits, such as supporting local food production and raising environmental awareness, it is crucial to consider its potential downsides on wild bee populations. Therefore, urban beekeepers need to operate responsibly, understanding wild bees' essential role in pollinating local flora. Increased education and awareness about the importance of wild bees and stronger regulations around urban beekeeping can help protect these critical contributors to our ecosystem.

CERTIFICATIONS

Certified Vegan: Run by the Vegan Awareness Foundation, the Certified Vegan logo indicates products are free from animal ingredients and animal testing, with adherence verified and certified.

The Vegan Trademark: Managed by The Vegan Society, the Vegan Trademark assures consumers that a product and its ingredients don't involve any animal use, including testing, during production.

5. Palm Oil Free

Palm oil use spells trouble for our planet! Palm oil often enters our consumption through various processed derivatives. These derivatives, numbering around five hundred, are often hidden under obscure names, making it challenging to identify palm oil products. To sidestep products containing palm oil, checking the ingredient labels for their various aliases is crucial. With hundreds of derivatives, identifying key letter combinations like PALM, STEAR, LAUR, and GLYC simplifies the process. Remember, spotting these terms in an ingredient doesn't automatically signify palm oil, but it's a lead worth pursuing with the company.

CERTIFICATIONS

The International Palm Oil Free Certification Trademark & Accreditation Programme: This global certification program authenticates products as completely palm oil-free.

Orangutan Alliance's label: The Orangutan Alliance is a nonprofit that awards a certification mark to products free from palm oil. This mark highlights palm oil-free products and supports conservation projects, particularly those assisting orangutans.

The Roundtable on Sustainable Palm Oil (RSPO): RSPO is a not-for-profit that developed a set of environmental and social criteria companies must comply with to produce Certified Sustainable Palm Oil (CSPO).

6. Better Packaging

Plastic stinks! The green beauty industry needs a serious reality check on its packaging options. Green beauty is about healthier ingredients, but what's the point if we're drowning in plastic? Sustainability isn't just about what goes on your skin but also what stays behind when the product is gone. Sustainable packaging refers to packaging with less environmental impact than traditional alternatives, a definition that may seem straightforward but carries broad implications. Sustainability spans the entire packaging lifecycle—from resource extraction and creation to usage (and reusability) and, ultimately, recycling or disposal.

Furthermore, sustainability encompasses economic and social dimensions. Even if a packaging solution is environmentally friendly, it might not be economically feasible for companies, making its continuous use unsustainable. Additionally, producing specific eco-friendly packages might have social repercussions, such as repurposing agricultural land for raw materials for bioplastics or causing deforestation on a large scale. Eco-friendly packaging involves several critical aspects. Firstly, it utilizes materials entirely recycled or derived from natural sources. Secondly, it requires efficient production processes, such as adopting smaller packaging formats, maintaining cleaner and well-orchestrated supply chains, and implementing zero-waste and closed-loop manufacturing models. Lastly, it promotes circularity by using packaging that can be easily repurposed, upcycled, or recycled.

The Environmental Protection Agency (EPA) reports that only 9.5 percent of plastic waste was recycled in 2014, 15 percent was incinerated for energy, and the remaining 75.5 percent was relegated to landfills. This illustrates the stark reality of plastic management, especially when considered in the context of cosmetics packaging.

Plastic is a popular choice for cosmetics packaging, particularly certain types, given their recyclability, energy-efficient production, and the ability to provide robust containers to safeguard the product through diverse conditions. High-Density Polyethylene (HDPE) is an excellent example of this, being BPA-free, widely recyclable, lightweight, and reusable multiple times. However, its utility diminishes with each reuse cycle, as "downcycling" reduces effectiveness and eventually necessitates disposal.

Plastic packaging, despite certain benefits, also carries significant disadvantages. Its non-infinite recyclability, origin from fossil fuels, and substantial environmental impact during extraction, processing, and disposal raise serious concerns. Additionally, plastics are not fully biodegradable, often contain potentially harmful chemicals like phthalates, and contribute to microplastic pollution.

Ideally, cosmetic packaging should use genuinely recycled plastic, free from BPA and phthalates, and be recyclable. It's even better if this plastic comes from waste, like discarded fishing nets.

Yet, it's important to note that not all plastic taken to recycling facilities is appropriately recycled. Factors such as inadequate cleaning, incorrect sorting, or mixing with wrong materials often lead to recyclable plastic being incinerated or dumped in landfills, producing toxic smog.

Considering all these considerations, HDPE or #2 plastic is a better choice for cosmetic packaging. It's commonly repurposed into toothbrushes, ornaments, plant pots, and children's toys.

But what about other types of materials? Let's break it down:

GLASS PACKAGING

- Made from abundant but nonrenewable materials like sand, soda ash, and limestone.
- Can be recycled, but the process is energy-intensive.
- Can be reused over and over.
- Weighs more than plastic, accounting for about a third of the total weight of a product.
- Can shatter during transport or freezing temperatures, leading to product waste.
- Ideally used in local bulk or zero-waste stores where containers can be refilled.
- If not managed properly, its environmental impact could outweigh its benefits.

PAPER PACKAGING

- Can be sourced from sustainable forests and made without chemicals.
- Recycling and composting of paper is easy.
- Resource and energy-intensive production; involves chlorine bleaching and other chemical processes that could create carcinogenic substances.
- For cosmetics, water-repellent chemicals are required, potentially rendering the container non-recyclable.
- Some manufacturers may overstate the percentage of recycled material used in their products.
- Overall, suitable for dry cosmetics that are mixed with water at home, but its use is limited due to its limitations.

METAL PACKAGING

- Aluminum, stainless steel, and tin are durable, attractive for recycling, safe, and function well in varying temperatures.
- Aluminum is resistant to corrosion and less energy-intensive to recycle than glass.
- Recycled aluminum requires significantly less energy during manufacture compared to new aluminum.
- However, metal packaging is opaque, not squeezable, and initially more expensive for manufacturers.
- Can be reused over and over.
- Tin and steel may have been lined with BPA, which raises concerns for daily use.
- Despite these drawbacks, BPA-free aluminum and stainless steel are excellent choices for a closed-loop packaging system.

POST-CONSUMER RECYCLED (PCR) MATERIALS

Unraveling the mystery of recycling labels can challenge many eco-conscious consumers. However, understanding the difference between products made from pre-consumer and post-consumer recycled content is crucial in making truly sustainable choices. Often, we gravitate toward anything bearing the "recycled" tag without fully comprehending its meaning, not knowing that the type of recycled content—pre-consumer or post-consumer—impacts its overall environmental benefit.

- **Pre-consumer recycled content** originates from manufacturing waste that was never used by consumers, such as scraps, rejects, or trimmings. These materials are repurposed instead of being discarded.
- **Post-consumer recycled content** is derived from materials consumers have used, disposed of, and kept out of landfills. Typical examples are aluminum cans and newspapers placed in recycling bins.

- Products labeled as simply **"recycled content"** could comprise either pre- or post-consumer waste or a mix of the two. Usually, a product containing a high proportion of post-consumer waste is specified as such.
- **Post-consumer content is more eco-friendly** than pre-consumer content. Manufacturers have always been adept at reusing pre-consumer waste, so it's less likely to end up in landfills. Conversely, post-consumer waste's environmental implications are higher because, if not appropriately recycled, it's more likely to clog landfills.

Although both pre- and post-consumer recycled content products are environmentally better than zero-recycled content products, the greener choice is to opt for post-consumer recycled content whenever possible.

BIODEGRADABLE PACKAGING

Biodegradable packaging, including materials like paper and certain types of bioplastics, has been positioned as an eco-friendly alternative to conventional plastics due to its ability to decompose naturally over time. However, despite this benefit, notable environmental concerns surround their use.

Although these biodegradable materials come from renewable sources, they often necessitate industrial composting conditions for efficient decomposition. These conditions are not typically achievable in household composting setups or the natural environment. Consequently, if these materials aren't disposed of properly, they might end up in landfills where they decompose no faster than conventional plastics.

The production of bioplastics also presents other environmental challenges. The cultivation of plant materials used in the production, such as corn or sugarcane, can strain land and water resources that could otherwise be utilized for food production. Additionally, the production process can contribute to pollution due to the use of fertilizers and pesticides and the chemical processing required to convert organic material into plastic.

Furthermore, a significant challenge lies in the fact that most recycling facilities currently do not have the infrastructure in place to efficiently process these types of biodegradable materials. This makes correct disposal and recycling a complicated issue.

Therefore, while biodegradable packaging offers certain advantages, it's crucial to consider these environmental concerns. We must improve the disposal and recycling infrastructure while minimizing the environmental impact during production.

OTHER OPTIONS

Dissolvable packaging refers to packaging that breaks down and disappears when exposed to certain conditions, such as water, heat, or light.

Refillable products or packaging have been introduced as a solution to waste, but they come with some challenges.

CERTIFICATIONS (FOR PAPER AND CARDBOARD PACKAGING)

The Forest Stewardship Council (FSC) is an international organization dedicated to promoting responsible management of the world's forests, ensuring that they are harvested in a way that preserves biological diversity and benefits the lives of local people.

The Sustainable Forestry Initiative (SFI) is a North American program focused on ensuring that forestry is practiced in an environmentally responsible, socially beneficial, and economically viable manner.

The Programme for the Endorsement of Forest Certification (PEFC) is a global alliance that promotes sustainable forest management through independent third-party certification, ensuring that wood and non-wood forest products are produced with respect for the highest ecological, social, and ethical standards.

Within the realm of green beauty, the focus is shifting. It's less about combating the issues that plague traditional models and more about providing a sustainable, long-term approach to skincare. Brands like **Plaine Products, Ethique, and Conscious Skincare** exemplify this shift. These brands advocate for eco-conscious practices by offering refillable options and products with minimal packaging.

But the path to green beauty isn't devoid of hurdles. Practical problems like the non-availability of refills at retail stores can limit accessibility for many consumers. Even when refills are available, their use can present challenges. For instance, refills might not fit perfectly into their original packaging, leading to functional difficulties and potential waste.

Moreover, while these brands push for sustainability, the allure of new products and formulas can overshadow the merits of refillable, repeat-use items. The market's rapid pace can lead to consumers preferring variety over sustainability, thereby questioning the practicality of refills in the first place.

Fortunately, more localized solutions are emerging to complement these green beauty initiatives. For example, zero-waste grocers or bulk shops encourage customers to bring their containers for refilling, reducing the need for new packaging.

While challenges exist in both conventional and green beauty sectors, the shift toward more sustainable practices continues to grow. The hope is that these challenges can be addressed effectively through sustained efforts and increased consumer awareness, leading to a sustainable beauty industry.

So how can we, as consumers, tackle this waste issue? How can we recycle or discard our empires responsibly?

In 2021, just 5–6 percent of the forty-six million tons of plastic waste generated in the US was recycled. And even when plastics are recycled, up to a third of the materials can be discarded due to contamination or process losses. When recycling beauty products, adhering to local recycling regulations is essential. Use resources like **Recycle Coach, How2Recycle**, and **EARTH911** to understand what materials are accepted in your area. Unfortunately, many beauty products don't meet the strict requirements of Material Recycling Facilities (MRFs). While the recycling triangle symbol may be on a product, it doesn't always mean the item is recyclable. Generally, only plastic items marked with the numbers 1 or 2 are widely accepted in curbside recycling programs.

The Detox Market stands out in the beauty industry with its robust sustainability initiatives. Through a partnership with Cleanhub, it addresses the issue of plastic waste, especially in the Global South. Collaborations with **TerraCycle** ensure the recycling of a significant percentage of collected waste while offering reuse and upcycle solutions for the remaining waste.

In addition to The Detox Market, several brands are launching recycling programs. **Nordstrom**, for example, has teamed up with TerraCycle to launch **BEAUTYCYCLE**, which facilitates recycling beauty and skincare product packaging via in-store collection points.

TerraCycle offers three different beauty product recycling programs that can handle various types of waste, such as mascara wands, pumps, lids, lipstick tubes, and mirrored compacts. **Colgate** sponsors an oral care program through **TerraCycle**, accepting any brand of oral care waste like toothbrushes and empty toothpaste or floss containers.

Many brands have established exclusive partnerships with **TerraCycle** for their recycling programs. These include **Burt's Bees**, **Glow Recipe**, **Supergoop**, **Paula's Choice**, and **Garnier**. Customers can collect a certain amount of empties from these brands, and TerraCycle will process them free of charge. Recycled plastics are often converted into useful items such as outdoor furniture, decking, storage bins, or playground equipment.

To incentivize recycling, some brands reward customers who return empty containers. For example, customers can drop off empties from any brand at **Credo Beauty's Pact Bin** and earn ten Credo Reward points per empty container. Similarly, **BareMinerals** and **Kiehl's** reward customers with points for recycling brand-specific containers in-store. Considering future sustainability, some brands have shifted from plastic to more recyclable materials. Brands like **Cocokind**, **Beautycounter**, **Pacifica**, **Osea**, and **Kiehl's** now use glass packaging for some or all of their products. To further cut down on packaging, brands like **Glossier**, **Kjaer Weis**, **Milk Makeup**, **Youth to the People**, and **Glow Recipe** have introduced refills for some of their popular products. These moves signify significant strides toward more sustainable consumption in the beauty industry.

Remember that items made from a single material are more likely to be recyclable when recycling beauty products. Combining different materials can make the recycling process costly and complicated. Items made from single materials such as glass, plastic, or cardboard can be rinsed straight into the correct recycling bin.

However, small items (under two inches) and dark packaging materials cannot usually be recycled due to technical limitations at MRFs. Other non-recyclable items include products that contain mirrors, magnets, makeup brushes, sheet masks and packets, and squeezable tubes.

In wrapping up, it's important to remember that "clean beauty" encompasses more than just the products themselves; it's also about a brand's dedication to both personal health and the well-being of our planet. The term "sustainable beauty" often holds a depth of meaning that may surpass the average consumer's expectations.

In a perfect world, regulations, and standards would make it effortless for everyone to understand the ethos and practices of the brands they support. However, given the current lack of regulations around using terms like "clean beauty" and "sustainable beauty," it falls on us to ensure that the brands we support align with our values and definition of clean beauty.

Regrettably, this isn't always as straightforward as it should be. As consumers, we shouldn't need to adopt the role of investigative reporters just to discern if a brand is genuinely sustainable or merely masquerading through greenwashing. So, as we navigate this evolving landscape of green beauty, let's remain diligent, informed, and, above all, committed to the health of both ourselves and our planet.

CHAPTER 7

Bathroom Revolution: From Waste-Land to Waste-Free

Embarking on a zero-waste journey may feel like standing at the base of a towering mountain—daunting. But every summit is reached by taking the first step. So, where do you begin? Where the rubber duck floats—your bathroom! Surprisingly, this intimate space offers the perfect setting for your initial stride into sustainability.

Like a treasure hunt, picture a joyous expedition into your bathroom cabinets and drawers. You're sifting through your trove of products, playing a game of "keep or toss," replacing the "toss" with healthier, environmentally friendly alternatives. It's a game show in your home with eco-friendly victories at stake!

Decluttering your bathroom not only frees up physical space but also introduces a sense of mental peace. As you systematically reduce, you're clearing out cabinets and shrinking your environmental footprint.

Indeed, adopting a zero-waste lifestyle might feel like learning a new language, where the grocery store transforms into your classroom, and reusable containers and compost become your vocabulary. But don't be daunted; I'm here to guide you through every product swap and eco-hack.

Despite their compact size, bathrooms are substantial contributors to household waste. They house numerous items requiring constant restocking—toothpaste, toothbrushes, floss, toilet paper, shampoos, and cleaning supplies. The challenge extends beyond the tangible plastic containers ending up in landfills or polluting our waterways. We must also acknowledge the less apparent forms of waste, such as the overuse of water, which constitutes a different type of wastefulness altogether.

Our bathrooms have unwittingly evolved into hubs of waste. It's time we pull back the shower curtain and confront these issues head-on, transitioning toward sustainable practices.

APPLY MY NINE Rs

Our journey into an eco-friendly bathroom transformation hinges on my comprehensive sustainability mantra: **The nine Rs**. These include Rethink, Refuse, Reduce, Reuse, Repair, Repurpose, Refill, Rot, and Recycle. Each one forms a significant pillar in constructing a sustainable bathroom.

Rethink and Refuse: These first steps call for a shift in mindset. Scrutinize your bathroom routine and purchases to discern between essentials and redundancies. Are there items you don't need or could refuse? For instance, if you see a new body scrub that catches your eye, rethink—could you make a similar one at home using natural ingredients?

Reduce, Reuse, and Repurpose: Strive to reduce waste by choosing sustainable alternatives for items nearing their end of life, like a bamboo toothbrush instead of a plastic one. Consider reusing items wherever possible. An empty glass jar, for instance, could be cleaned and reused to store homemade lotions or soaps. If things can't be reused as they are, get innovative. A worn-out bath towel, for instance, could be repurposed into a bathroom rug.

Repair and Refill: Before discarding broken items, explore options to repair them. Does your electric toothbrush need a simple battery change, or can your hairdryer be fixed by replacing a small part? Choose to refill products where possible. Buying bulk soap or shampoo and refilling your containers reduces packaging waste and proves cost-effective.

Rot and Recycle: Create a compost system for organic waste, such as bamboo toothbrushes, hair, and nail clippings. For non-compostable items, have a dedicated recycling bin in your bathroom for materials like hard plastic, glass, metal, paper, and cardboard. Remember to rinse containers before recycling and check if items like pumps or spray tops can be recycled.

SUSTAINABLE SWAPS

1. Toilet Paper

The environmental impact of toilet paper production is immense and often overlooked. Each year in the US, toilet paper production consumes up to fifty-four million trees, requires massive amounts of electricity, and uses around 474 billion gallons of water. As the use of toilet paper increases globally, these environmental costs continue to rise.

Brands like Charmin, Cottonelle, and Quilted Northern, made entirely from virgin forest fiber, rank low in sustainability. The "tree to toilet" pipeline associated with these brands significantly contributes to the degradation of the climate-critical Canadian boreal forest. Each year, the tissue industry clears one million acres of this forest, positioning Canada third in global intact forest loss rankings, trailing only Russia and Brazil.

These forests aren't just trees; they're crucial carbon sinks, with their vegetation and slowly decaying soils containing nearly twice the carbon in the world's recoverable oil reserves. The forests are also home to unique ecosystems and Indigenous communities whose livelihoods and ways of life are threatened. We're accelerating toward a climate catastrophe with every flush of virgin forest fiber toilet paper.

Yet, more sustainable alternatives to traditional toilet paper are available. Brands like **Who Gives A Crap, Reel, and PlantPaper** offer eco-friendly toilet paper from 100 percent recycled materials or sustainably sourced bamboo. Not only are these products less harmful to the environment, but their packaging is also compostable, reducing landfill waste.

An even more sustainable solution is the use of bidets. Companies like **Hello Tushy** and **Whisper Bidets** offer bidet attachments that can be added to your existing toilet setup, providing a near-zero-waste option for personal hygiene. TMI—I can't live without mine!

Contrary to the assertion that bidets waste water, the amount utilized is relatively tiny compared to what's required for toilet paper production. Specifically, a high-end bidet uses roughly an eighth of a gallon of water, while a standard toilet uses about four gallons per flush. In contrast, creating a single roll of toilet paper consumes 37 gallons of water, 1.3 kilowatt/hours of electricity, and around 1.5 pounds of wood.

The debate around reusable toilet paper (yes, there is such a thing) or "family cloth" has sparked controversy. Proponents argue it's eco-friendly and cost-effective, while critics point to increased laundry, potential hygiene issues, and the awkwardness of offering cloth wipes to guests. While reusable toilet paper can reduce the waste of traditional toilet paper, it might not provide the best balance of sustainability, cost-effectiveness, and convenience. I must admit I am not ready for this swap and might never be!

2. Soap/Body Wash

Transitioning to zero-waste soap typically eliminates liquid hand and body soaps in plastic pump bottles, particularly those containing plastic beads. Bar soaps have existed for centuries and serve as easy-to-use, zero-waste alternatives.

You can often find such soaps at local health food stores, which often display a variety of package-free soaps. All you need to do is check the ingredients list for undesirable concoctions and bring your reusable container for storage.

Online shops like **etee** offer hand and body soap bars devoid of plastic packaging.

Companies like **Ethique** offer sustainable soap bars in compostable packaging.

There are still sustainable options if you're not ready to release your liquid soaps. Some companies, like **Plaine Products** and **Public Goods**, offer body wash refill programs.

3. Shampoo and Conditioner Bars

Lather, rinse, repeat? How about lathering, rinsing, reducing waste, and repeating?

These environmentally friendly products help reduce plastic waste with their compostable or biodegradable packaging. Moreover, they often contain natural and organic ingredients that are kinder to your hair and skin than the harsh chemicals in conventional products. They're also cost-effective, as they're typically concentrated and long-lasting. These compact, spill-proof bars are perfect for travel, and there's a wide range available to suit different hair types.

A selection of brands are demonstrating their commitment to both sustainability and quality in the zero-waste shampoo and conditioner market. **etee**, for instance, provides various plastic-free shampoo and conditioner bars formulated with natural ingredients. **EcoRoots** offers a range of environmentally friendly, chemical-free shampoo and conditioner bars packaged minimally in recyclable materials. **UpFront Cosmetics** is another key player, producing natural, compostable hair care products. **The Eco Alchemis**t takes a holistic environmental approach, offering solid hair care options in reusable, recyclable tins, aligning their products and packaging with a firm waste-reduction pledge.

Everist stands out for their creative approach to packaging. They have moved away from the traditional water-based formula found in most shampoos and conditioners, opting for a waterless, highly concentrated product. The resulting smaller, lighter packaging reduces shipping resources and is also made of aluminum, which is endlessly recyclable.

Lastly, **Meow Meow Tweet** has taken an innovative method to shampoo and conditioner, offering them in powder form. All you need to do is add water. Packaged in a metal bottle, this is another brilliant and easy way to reduce plastic waste in your hair care routine. The bottle can be refilled, repurposed, or recycled, adding to the product's sustainability credentials.

When you transition from conventional shampoos to shampoo bars, be ready for some changes due to the absence of synthetic detergents and harsh chemicals that strip natural oils from your hair. This shift can initially make your hair look greasier as it produces the same amount of natural oils without removing them. This adjustment period, often the first two weeks, is temporary. With patience and an open mind, you'll find the right shampoo bar among the many available, each offering unique benefits.

4. Deodorant

Plastic deodorant sticks, while convenient, present significant environmental issues. These products often come in plastic containers with various components, making recycling difficult. Most people don't dismantle these containers for recycling, and many end up in landfills.

As an alternative, consider options that come in recyclable or compostable packaging. For instance, **Piperwai**, a BIPOC-owned brand, creates effective deodorants suitable for sensitive skin packaged using Ocean Waste Plastic, which can be recycled by most curbside programs.

Other sustainable alternatives come in pastes, such as those from **I Luv It** and **No Pong**. **Ethique** has pioneered naked deodorant bars, eliminating packaging waste entirely, while others choose to use compostable cardboard tubes for their products.

When considering a switch from conventional to natural deodorants, be aware of the differences in formulation. Traditional aluminum-based deodorants work as antiperspirants, blocking sweat pores. On the other hand, natural deodorants often use ingredients like baking soda, arrowroot powder, clay, or charcoal to absorb moisture. Instead of preventing sweat, they quickly soak up sweat and neutralize or mask odor. However, too much baking soda can cause skin irritation. Many natural deodorants are now offering baking soda-free options.

Remember that natural deodorants' effectiveness can vary from person to person. Personal body chemistry, chosen scent, and application form can all impact their performance. Experimenting may be necessary to find the right fit. Natural deodorants come in various forms, including sprays, powders, putties, creams, gels, sticks, and more.

5. Toothcare

TOOTHPASTE

According to Forbes, approximately a billion tubes are in landfills annually. The complex composition of toothpaste tubes—a mix of various plastics, aluminum, steel, and nylon—makes them challenging to recycle, as each material requires different recycling processes.

Beyond the plastic problem, conventional toothpaste brands often contain ingredients harmful to our health and the environment. These include triclosan, carrageenan, sodium lauryl sulfate (SLS), propylene glycol, artificial colors, and sweeteners. Triclosan, for instance, is not only a suspected thyroid disruptor that could contribute to antibiotic resistance, but it's also highly toxic to aquatic life and can disrupt animal development.

Fluoride, a mineral, is widely used in toothpaste due to its ability to fight tooth decay. However, it's essential to manage fluoride use in children properly. Young children, especially those under three years old, should use fluoride toothpaste roughly the size of a grain of rice. Children aged three to six should use a pea-sized amount. This precautionary measure is because children may swallow toothpaste, and excess fluoride ingestion can lead to dental fluorosis, resulting in tooth discoloration and damage. Although some research links high fluoride levels to cognitive problems in children and thyroid issues, these concerns typically arise from fluoride levels much higher than those found in toothpaste, and the evidence is inconclusive.

More and more companies are offering innovative and sustainable alternatives to conventional toothpaste. These come in various forms, such as toothpaste tablets, powders, and pastes. Toothpaste tablets, which dissolve in your mouth like mints, allow for precise dosage and eliminate the need for plastic tubes. Powders offer another environmentally friendly option, typically consisting of natural ingredients that you can apply using a damp toothbrush. Like traditional toothpaste, pastes are now offered in eco-friendly containers like glass jars and aluminum tubes.

Unpaste offers eco-friendly tablets in compostable wrappers. They offer options without fluoride and various flavors, including mint and cinnamon.

Bite delivers toothpaste bits housed in recyclable or upcyclable glass containers.

Hey Humans presents a natural, budget-friendly toothpaste packaged in recyclable aluminum tubes.

Change is a Canadian brand offering sustainable toothpaste tablets in compostable pouches.

David's toothpaste employs a refillable tube system with a metal key to help roll the tube and squeeze out the paste.

All of these brands offer both fluoride and fluoride-free options.

TOOTHBRUSHES

Plastic toothbrushes contribute significantly to plastic waste, with one billion discarded annually in the US alone. These toothbrushes, made from fossil fuel-derived polypropylene plastic and nylon, often end up in waterways and oceans, posing severe threats to marine life. An average person uses three hundred toothbrushes in their lifetime.

The challenge of recycling toothbrushes arises from their composition of nylon bristles, metal staples, and handles made from plastic or compostable bamboo. These components require disassembly for efficient recycling—a step most people are unlikely to undertake.

Instead of plastic, Bamboo brushes are considered sustainable due to their high growth rate and low environmental impact. It can grow to a meter in a single day and does not require replanting after harvesting as it regrows from its roots. Moreover, it produces 35 percent more oxygen than an equivalent stand of trees and absorbs high amounts of CO_2, thus contributing positively to climate change mitigation. Its natural antibacterial, antifungal, and antimicrobial properties make it an ideal material for oral care products. Bamboo products are 100 percent biodegradable and can be composted after use, reducing landfill waste.

With the explosion of bamboo products a few years ago, we have seen abundant greenwashing in this space. While bamboo might be a sustainable crop, in many cases, how it's sourced and manufactured is not.

Nylon, the most common material used for toothbrush bristles, is not sustainable as it is a petroleum-based product. Nylon is not biodegradable, contributing to landfill waste and environmental pollution.

Biobased nylon, while a step in the right direction, still poses challenges. It's made from plant-based materials instead of fossil fuels, reducing its carbon footprint. However, it's only biodegradable under specific conditions, which are hard to achieve in standard composting or recycling processes. Therefore, while biobased nylon is better than traditional nylon, it's not a perfect solution.

Some companies are exploring innovative ways to make toothbrush bristles more sustainable. One example is **Brush with Bamboo**, whose toothbrushes are certified as Biobased Products by the US government's Biopreferred Program. The bristles of these toothbrushes are made from 100 percent Castor Bean Oil and contain no fossil fuels or petroleum. Although not legally classified as "compostable" due to their slower degradation rate, these biobased bristles will eventually return to the soil, with the exact timeframe dependent on soil or ocean activity.

However, it's crucial to scrutinize claims of "biodegradable bristles." Always ask for testing paperwork supporting such claims and beware of bristles made from PLA (Polylactic Acid). Derived from genetically modified corn, PLA breaks down too quickly for repeated use in the mouth.

Alternatively, some brands produce toothbrushes with bristles made from boar hair, offering a fully compostable option. Yet, the requirement to slaughter pigs for their hair raises ethical considerations.

Brand recommendations: Georganics, The Humble Co., and Brush Naked.

MOUTHWASH

Avoid the conventional stuff. It has a mouthful of ingredients I can't pronounce, most notably coal tar and alcohol.

Brand recommendations: By Human Kind's Mouthwash Tablets, Nix's Refillable Mouthwash, and Georganics Oil Pulling Mouthwash.

FLOSS

Common Toxins Found in Dental Floss

PFAS (Per- and Polyfluoroalkyl Substances): These artificial chemicals, widely used in everyday products since the 1950s, now contaminate our water supplies and are found in 99 percent of Americans due to their prevalence. They've been associated with various health issues, including liver, kidney, and immune system damage, thyroid disease, cancer, hormonal imbalances, and developmental delays in children. Alarmingly, these "forever chemicals" do not degrade and can persist for thousands of years in our bodies and the environment.

Plastic: Many dental floss brands utilize plastic and petroleum derivatives, such as nylon and polyester, which often contain toxic endocrine disruptors like BPA. These disruptors have been linked to several health problems, including infertility and thyroid dysfunction. Furthermore, the plastic in dental floss contributes to environmental pollution as it cannot be recycled and annually adds to the eighteen billion pounds of plastic dumped into the ocean.

Artificial flavors: Traditional floss uses artificial flavors to create a minty taste. However, the safety of these flavors is still debated due to inadequate regulation in the US and insufficient long-term safety research. The FDA has removed seven synthetic flavoring substances from its approved list. The lack of transparency about the components of artificial flavors could lead to consumers unknowingly exposing themselves to inflammatory ingredients when using flavored floss.

Alternatives

Silk is a natural, biodegradable, gentle option for gums. Peace Silk is a good alternative for those concerned about the ethical implications of traditional silk production. It involves a process that doesn't harm the silkworms during their metamorphosis.

Bamboo: As a plant-based, vegan option, bamboo floss is advantageous as it grows abundantly without the need for toxic pesticides, and it decomposes rapidly. However, transforming bamboo into a flexible product can require many chemicals, so it's essential to consider the manufacturing methods when choosing bamboo floss.

Corn fiber: Also suitable for vegans and those following a plant-based lifestyle, dental floss made from corn fiber (often labeled as corn PLA) is a natural, biodegradable, and highly compostable alternative. It's recommended to seek out corn fibers sourced from non-GMO corn for a more eco-friendly choice.

Brand recommendations: etee uses pure mulberry silk. **Public Goods** offers biodegradable silk options. **EcoRoots** uses corn PLA and candelilla plant vegan wax. **Geoorganics** is made from activated charcoal, corn PLA, and vegetable wax. All of them are packed in glass bottles and some offer refills.

Water Pik Flosser

This is another good option. You'd only buy this once. It's not zero waste, but I wanted to include it as another option for flossing. Water flossing has many benefits. It's electronic and uses a strong jet stream to clean in between the teeth.

DIY Flossing

Many years ago, I attended a talk with Bea Johnson, and the one thing that stood out to me was her DIY floss idea. She recommended you unravel a piece of silk fabric and use that to floss with. I did try this and found I had to be gentle as it broke easily. I do suggest you give it a go, though! Everyone's teeth and spaces are different.

5. Razors

158.10 million Americans used disposable razors in 2020, projected to increase to 160.16 million in 2024 The main problem is that disposable razors can't be recycled because they are made of plastic, metal, and rubber. All those parts need to be detached, and most recycling facilities do not have that technology.

Safety razors are made using long-lasting stainless steel or chrome. They are entirely plastic-free and have a firm handle and double-sided blades. You can remove the whole blade to replace it, making it more sustainable than its plastic counterpart, where you must toss the entire head. You have two options from the standard: remove the blade by screwing off the whole blade. In the butterfly version, you have to twist to open the blade chamber to remove the blades.

Schick and Gillette, well-known razor manufacturers, are taking steps toward sustainability. Schick practices zero-landfill production and provides a razor recycling program. Similarly, Gillette's razors are now recyclable across the US. However, recycling rates remain low, contributing to the problem of disposable plastic razor handles filling up landfills.

Switching to more sustainable options can also be cost-effective. For example, a disposable Venus razor can cost more than $15, while a pack of ten blades from **EcoRoots** is priced around $4.97. That's a huge saving.

Several brands, such as **Leaf Shave**, **Well Kept**, and **Jungle Culture**, offer eco-friendly alternatives. If you're looking to reduce waste even further, consider alternatives to shaving. You could embrace your natural hair or try waxing, preferably sugar waxing, due to its lower waste output. An epilator, which removes hair at the root, is another viable option.

When disposing of used blades, it's crucial not to throw them in the trash wrapped in toilet paper, as this can pose a danger to waste management workers and animals scavenging in landfills. Instead, store used blades safely until you can dispose of them correctly. Check with your local municipality for guidelines on handling metal waste. You can also send used blades of any brand to Leaf Shave for appropriate disposal.

6. Makeup

In the previous chapter, I discussed the beauty industry's obsession with plastic and toxins. Imagine if every time you put on your favorite shade of lipstick, you also painted a one-inch plastic tube. Now, think about that cheeky blush that adds color to your face. Imagine it also adding a handful of plastic pebbles into our oceans. Each coat of your volumizing mascara could be seen as brushing a tiny plastic twig onto our forests.

How about your foundation? Every pump of that liquid perfection might as well be spreading a thin layer of plastic film over our precious soil. And don't forget the eye shadows—each dash of vibrant color could equate to sprinkling plastic dust into our air!

It's not just your makeup routine; it's a global affair. Imagine the sheer volume of plastic these beauty essentials are dressed up in daily worldwide. That's a beauty pageant no one wants to win!

BRAND RECOMMENDATIONS:
(THERE ARE SO MANY; HERE ARE A FEW FAVES)

Izzy Zero Waste Beauty sets a new standard for sustainability, with their products shipped in reusable mailers created from upcycled materials. Their containers are crafted from medical-grade stainless steel, and the best part? When you're done with your product, you can return the empties, which will be cleaned and refilled.

Elate Cosmetics' operations are roughly 75 percent waste-free, employing bamboo, glass, aluminum, and seed paper packaging with limited recyclable plastic. Their approach proves that beauty doesn't need to come at the expense of the environment.

Ilia stands out with its take-back program in partnership with TerraCycle. US customers are encouraged to mail up to ten empty beauty products per month (from any brand) for recycling, taking a step toward zero waste.

Sappho New Paradigm offers refillable compacts to hold magnetic pans that can be easily swapped, promoting reusability.

Clean Faced Cosmetics echoes a similar ethos, with an option to order refills or return empties for reuse, preventing unnecessary waste.

Fat and the Moon ensures their products come in reusable, recyclable containers, emphasizing the value of reusability in their brand mission.

Axiology goes further with its lipstick tubes made from 50 percent recycled plastic. Their red boxes are handmade from recycled trash, while Balmies and highlighters come in recyclable paper. They also have a lipstick tube mail-back system.

7. Cotton Pads and Wipes

Disposable makeup wipes are a primary environmental concern. They contribute significantly to landfill waste as they are single-use items. Millions of makeup wipes are discarded every day, and, unlike biodegradable waste, these wipes don't decompose easily, lingering in the environment for years and adding to our already overwhelming waste problem.

Many makeup wipes are made of nonrenewable resources such as polyester, a petroleum-derived plastic. This material is not biodegradable and can take hundreds of years to break down in the environment. Furthermore, the manufacturing process of these polyester-based wipes contributes to pollution and requires a substantial amount of energy.

Makeup wipes are often packaged in non-recyclable plastic, further exacerbating plastic pollution. They also require regular repurchasing, leading to a continuous cycle of consumption and disposal.

Conventional cotton pads pose several environmental issues. Firstly, the cultivation of cotton is extremely water-intensive. It's estimated that it takes over 5,000 gallons of water to produce a mere kilogram of cotton, contributing to water scarcity in many regions. Moreover, cotton farming heavily relies on pesticides and synthetic fertilizers, which can contaminate local water supplies, harm wildlife, and pose health risks to farm workers.

Brand recommendations: Tru Earth Bamboo Rounds Reusable Makeup Remover Pads, Oko Creations Hemp and Organic Bamboo Rounds, Elva's Bamboo Pads, and Conscious Skincare Organic Cotton Reusable Makeup Pads.

So, you've heard of these so-called "flushable" wipes, right? They seem green and convenient. But let me spill the beans. It's a case of greenwashing at its finest. See, unlike our trusty old friend toilet paper. These wipes don't disintegrate quickly. Nope, they hold it together like champs, often for months. The result? A high potential for clogging up your plumbing system, big time.

And, you know what's worse? They get together with fats, grease, and other undesirables in the pipes, forming these massive, troublesome clusters called "fatbergs." We're talking ten-foot-long, hundred-pound monsters causing significant blockages.

Let's not forget our older houses with plumbing systems that are, let's say, past their prime. The last thing these old pipes need is a wipe-induced clog. Yet, if we're flushing these wipes, that's precisely what they might get. Roots can infiltrate the system, creating a web inside the pipes that traps anything, especially tough wipes. Talk about a plumbing nightmare!

And suppose these pesky wipes manage to reach your home's septic system or city sewer. In that case, they continue their destructive journey, causing blockages and forcing sewage back into your home. Yikes! Besides the gross-out factor, we're discussing some costly clean-ups and repairs.

Even when they reach sewage treatment plants, these wipes don't quit. They can clog up the big industrial pumps there, causing them to break down or even burn out. And who ends up footing the bill for that? Yep, we taxpayers!

So, the moral of the story? No matter how "flushable" they say they are, do your plumbing (and wallet) a favor—bin those wipes, don't flush them, or even better, opt for reusables made from organic cotton or bamboo!

8. Period Care

The sheer volume of waste associated with menstruation is staggering. An individual who menstruates is estimated to use 5,000 to 15,000 pads and tampons over their lifetime, resulting in approximately four hundred pounds of packaging waste alone. Twenty billion disposable menstrual products are discarded annually in the US, generating 240,000 tons of solid waste. But the impact extends beyond mere numbers.

Navigating through the effects of the combination of toxic contaminants in disposable menstrual products requires a significant shift in how we perceive period care.

TAMPONS & PADS

To say that this is a complicated subject is an understatement; I could write a whole book on this subject. But here's the gist of it.

Conventional tampons are typically made from a blend of cotton and rayon, materials recognized for their high absorbency yet are environmentally unsustainable. Traditional cotton, often used in these products, is grown using nonorganic pesticides and is considered one of the world's "dirtiest" crops. Rayon, a semi-synthetic fabric, undergoes a production process involving various chemical additives that ideally should not come into contact with sensitive areas of the body.

These tampons, already riddled with chemicals, are treated with additional substances during manufacturing, including chlorine bleach and fragrances. The tampons are then packaged in a plastic applicator containing Bisphenol A (BPA), an endocrine disruptor associated with infertility.

A cocktail of chemicals, including dioxins, methylene chloride, glyphosates, carbon disulfide, chloroform, styrene, and synthetic fragrances, have been detected in traditional tampons. These chemicals are potential irritants, carcinogens, and reproductive toxins. Since tampons are worn for several hours, these substances can leach directly into the body through the absorbent vaginal mucosa and enter the bloodstream.

Opting for organic tampons manufactured from chemical-free materials minimizes exposure to these harmful substances. However, it's important to note that organic tampons do not reduce the risk of Toxic Shock Syndrome (TSS), a condition caused by bacterial infection. To mitigate TSS risk, select the appropriate tampon absorbency, change your tampon regularly, and ensure good hand hygiene.

Traditional versions of sanitary pads may appear and feel like cotton, but they're usually crafted from synthetic plastics such as rayon, polyester, or superabsorbent polymers (SAPs). These materials are not biodegradable and contribute to landfill overflow and greenhouse gas emissions. The production process of these pads also relies heavily on fossil fuels.

In addition, toxins found in traditional pads have been linked to several health effects:

- **Dioxin:** Linked to hormonal disruption, cancer, and endometriosis, even at trace levels.

- **Pesticides:** Used in GMO cotton production, they're associated with neurological dysfunction, infertility, and developmental defects.

- **Fragrances:** Often undisclosed due to "trade secrets," they often contain toxic chemicals.

- **PFAS:** Known as "forever chemicals," they're associated with various health problems, including decreased fertility, high blood pressure in pregnant individuals, increased risk of certain cancers, and more.

- **Rayon:** So toxic that its manufacturing is banned in the US.

- **Glyphosates:** A known carcinogen found in 80 percent of sanitary products.
- **Titanium dioxide:** A pigment that could potentially cause cancer.
- **BPA:** Found in most plastics, it can interfere with the reproductive system and contribute to cancer.
- **Chlorine bleach:** A potential carcinogen and irritant that gives pads a clean, white look.

Indeed, the oversight of feminine hygiene products by regulatory bodies like the US Food and Drug Administration (FDA) and the European Commission is essential to this discussion.

The FDA, responsible for protecting public health in the United States, classifies tampons as medical devices. Consequently, they oversee the safety and effectiveness of these products. While many chemicals found in tampons, such as dioxins, are FDA-approved, it is essential to note that the agency generally considers them safe based on their low levels of exposure. The FDA maintains that the trace amounts of dioxins in tampons resulting from the bleaching process are safe, despite concerns raised by some research and advocacy groups about the long-term exposure risks.

The agency also requires manufacturers to provide detailed labeling instructions, including information about the risk of Toxic Shock Syndrome (TSS). However, it does not mandate companies to disclose the complete list of ingredients in tampons because they are categorized as medical devices. This controversial issue has led to public demands for more transparency.

The European Commission, the executive branch of the European Union, takes a slightly different approach. Tampons and sanitary pads in Europe must comply with the EU's General Product Safety Directive, which requires that all products sold be safe for consumers. This body has stricter regulations regarding chemicals in consumer products, which extends to feminine hygiene products. For instance, they have banned certain phthalates, chemicals often used in plastics, which the FDA has only restricted in some products.

In recent years, the European Commission has also been focusing on the environmental impact of sanitary products. It has started encouraging member states to promote the production and use of sustainable menstrual products that are reusable or made from organic and biodegradable materials.

Both agencies play critical roles in overseeing the safety of feminine hygiene products. Their standards and regulations aim to balance consumer safety with the practicalities of manufacturing and commerce. However, advocacy for greater transparency and stricter regulation of these products continues as public awareness and concern about potential health risks grow.

There are various alternatives to traditional tampons and pads, each with unique benefits:

Reusable pads: These are typically made from cotton and can be washed and reused for several years. The key benefit is that they're environmentally friendly, helping to reduce the waste generated by disposable menstrual products. They're also free from traditional pads' chemicals and synthetic materials, potentially reducing the risk of irritation and other adverse health effects.

Recommended brands: Rael, Tampon Tribe, and Smartliners

Period panties: Underwear designed to absorb menstrual blood. They can be worn alone or as backup protection on heavier days.

Recommended brands: The Period Company, ModiBodi, and Aisle

Menstrual Cups: Small, flexible cups made of silicone, rubber, latex, or elastomer inserted into the vagina to collect menstrual blood. They can be worn for up to twelve hours before emptying, cleaning, and reinserting. Menstrual cups are cost-effective over time and reduce waste since they're reusable. They're also free from chemicals found in tampons and pads.

Recommended brands: Cora, Saalt, Pixie, and Diva

Menstrual discs: Like menstrual cups, menstrual discs are inserted into the vagina to collect blood. However, they're often thinner and broader and sit at the base of the cervix rather than in the vaginal canal. One of the key benefits is that they can be worn during sex, unlike most other menstrual products.

Recommended brands: Nixit and Intima

Organic tampons and pads: These products are made from organic cotton, free from pesticides and other harmful chemicals used in traditional tampons and pads. They're often biodegradable, reducing their environmental impact. The main benefit is reduced exposure to potentially harmful chemicals, although they still contribute to waste as they're usually not reusable.

Recommended brands: Natracare & Lola

When choosing personal hygiene products such as tampons, pads, period underwear, and menstrual cups, paying close attention to the materials and certifications that come with these items is essential. The safety and sustainability of the materials in these products are paramount for your health, as well as for the health of our planet. Certifications can assure these qualities, as they validate that the products meet specific standards related to organic farming, ethical sourcing, and ecological impact.

Tampons

100 percent certified organic cotton: Cotton is one of the most pesticide-intensive crops grown. Organic cotton is grown without toxic pesticides or synthetic fertilizers, making it safer for the environment, farmers, and users. It's hypoallergenic and breathable than synthetic materials, making it ideal for intimate hygiene products like tampons.

Certifications: Look for tampons with certifications such as OEKO-TEX, Global Organic Textile Standard (GOTS), Italian Association for Organic Agriculture (ICEA), Biodegradable Products Institute (BPI), Forest Stewardship Council (FSC), Ecocert, and Organic Content Standard

(OCS). These certifications assure that organic cotton is genuinely organic, free from harmful substances and that environmental and social responsibilities are considered in its production.

Pads

100 percent certified organic cotton and bamboo: Like tampons, organic cotton is an excellent pad choice. Organic bamboo is another sustainable choice. Bamboo grows quickly, requires less water than cotton, and has natural antimicrobial properties.

Certifications: In addition to the certifications mentioned for tampons, pads can also be USDA Biopreferred and MADE SAFE certified. USDA Biopreferred refers to products derived from plants and other renewable agricultural, marine, and forestry materials. MADE SAFE certifies that products are made with safe ingredients, not known or suspected to harm human health or ecosystems.

Period Underwear

Fabrics: While GOTS organic cotton period panties are still a rarity due to the complex nature of making effective and sanitary absorbent layers, some brands are making nontoxic period underwear with PFAS and PFOA-free synthetics, wool, hemp, bamboo, lyocell, or modal fabric.

Certifications: Look for period underwear with Better Cotton Initiative (BCI) and OEKO-TEX certifications. BCI promotes better standards in cotton farming and practices across twenty-one countries.

Menstrual Cups

100 percent medical-grade silicone: The safest material for menstrual cups is 100 percent medical-grade silicone, free from plastic fillers, BPA, and other bisphenols, as well as latex. Most of the cups on the market use this material. Note that "organic" menstrual cups don't exist because organic silicone doesn't exist. However, medical-grade silicone is considered safe and hypoallergenic.

9. Shower Curtains

Polyvinyl chloride, commonly known as PVC, is widely used in everyday products, including shower curtains. However, this plastic is associated with significant health and environmental risks due to its composition and the toxins it releases, such as chlorine by-products (dioxins and furans), phthalates, and volatile organic compounds (VOCs). Exposure to these toxins can lead to various health issues, including cancer, reproductive and developmental problems, hormonal disturbances, and damage to the immune system. PVC's toxicity can also be exacerbated by heat, such as during a hot shower.

To mitigate these risks, some consumers are turning to polyethylene vinyl acetate (PEVA) shower curtains as an alternative. Although PEVA is marketed as an environmentally friendly and nontoxic option, it's important to note that it still contains VOCs, which could be potentially harmful. However, PEVA is considered safer than PVC because it does not contain chlorine and does not release highly toxic dioxins.

Consumers are advised to opt for shower curtains made from natural, organic materials instead, especially for children's bathrooms.

PEVA's eco-friendliness is also disputed, as it is unclear whether it is biodegradable. It is more recyclable than PVC, but local recycling capacities may vary.

Materials like hemp, linen, and cotton are recommended in nontoxic shower curtains. Hemp is an excellent choice due to its natural resistance to bacteria and mold, durability, and ability to grow without harmful chemicals. It is machine washable and, although not waterproof, is effectively water-resistant when woven tightly. Linen offers similar benefits to hemp, with added anti-static properties. Organic cotton is preferable to conventional cotton, as it is grown and processed without toxic chemicals and uses less water. However, neither cotton type is mold-resistant or waterproof.

Other materials like nylon and polyester are better than PVC but remain less eco-friendly than natural alternatives, as they are still plastic and can contain VOCs. Whether you need a shower curtain liner largely depends on the curtain material, the type of weave, and personal preference. Some people choose fabric curtain liners to reduce their use of plastics.

If you decide to use a natural fabric shower curtain, you can waterproof it with Rawganique's Waterproofing Wax, made from natural ingredients free of petroleum derivatives, distillates, and chemicals.

Beyond materials, when buying a shower curtain, look for nontoxic dyes, third-party certifications (such as GOTS, OEKO-TEX, and MADE SAFE), ethical supply chains, and other sustainable initiatives. Remember, organic certifications can be expensive for small farmers, and the absence of such certificates doesn't always mean the product is non-eco-friendly.

Brand recommendations: Parachute, Rawganique, Dream Designs, and Coyuchi

10. Water Consumption

There are many ways to reduce water use around the house, and nowhere is this more critical than the bathroom. And we're not just talking about turning off the tap when brushing your teeth, though that's important too. But it's also about how your bathroom is designed.

Thinking about saving water shouldn't just be for when there's a drought or restrictions. It's a big part of how we impact the environment and something we should always remember, wherever we live.

Regarding the shower, it's not just about picking a water-saving showerhead—the controls are just as important. Picture this: a top-notch thermostatic shower that balances the hot- and cold-water supply. It responds to changes in water pressure or temperature, so if someone flushes the toilet or runs the bath elsewhere in your house, your shower stays at the same temp. This means less water and energy wasted.

There are also these cool eco-controls you can incorporate into your shower room. Think of hand showers with a water-saving button to control the water flow, dialing down consumption and letting you tweak the shower spray to your liking.

Now, let's chat toilets, the biggest water-guzzlers in the bathroom. A simple way to save water here is the good old dual-flush system. With households flushing away nearly six gallons of water daily, an efficient toilet can make a big difference.

Remember, saving water is about more than just saving money (though it does that too). It's also about reducing energy use and wastewater, which is better for the planet.

As for showerheads, the right choice can impact how much water you're using. Swapping them out can be a simple bathroom tweak. You've got these innovative spray technologies that can combine water-saving features with a luxury showering experience. Imagine water-efficient showerheads that mix air and water to create a powerful spray that gives any power shower a run for its money.

One last tip: Check out a shower's flow rate to see how much water it uses. If you're after an eco-friendly shower, you should see a flow rate of about 2.5 gallons per minute, less than the standard three to four gallons per minute.

OTHER WAYS TO TACKLE YOUR WATER WOES

Check for toilet leaks. Try this: add a bit of food coloring to your toilet tank. If you see color seeping into the bowl without flushing, then you've got a leak. It's something you'll want to get fixed pronto.

Don't treat your toilet like a trash can or ashtray. If you're flushing away cigarette butts or tissues, you're wasting about five to seven gallons of water each time. Let's reserve the toilet for its actual purpose, okay?

Use plastic bottles in your toilet tank. This is a cool hack. Take two plastic bottles, add a bit of sand or pebbles to weigh them down, fill them up with water, and put them in your toilet tank (just make sure they're not interfering with the moving parts). This little trick can save you ten or more gallons of water daily in a typical home.

Keep an eye on your faucets and pipes for leaks. It's amazing how much water a tiny drip can waste—we're talking twenty or more gallons daily. And if you've got a bigger leak, that could be hundreds of gallons.

WAYS TO MAKE WATER HEATERS MORE ENVIRONMENTALLY FRIENDLY

1. **Insulate the heater:** Insulating your hot water heater can significantly reduce heat loss and save energy. It helps to maintain the temperature of the water in the heater for a longer period, reducing the energy required to heat water.

2. **Regular maintenance:** Ensure regular maintenance of the heater to keep it operating efficiently. This includes periodic draining to remove sediment and checking for leaks or other issues.

3. **Temperature adjustment:** Lower the thermostat setting on your water heater to around 120°F (49°C). For every 10°F reduction in temperature, you can save between 3 and 5 percent in energy costs.

4. **Upgrade to a tankless or on-demand heater:** Tankless or on-demand heaters heat water only when required and thus save energy as compared to traditional heaters that maintain the temperature of the stored water.

5. **Invest in a heat pump water heater:** These heaters use electricity to move heat from the air or ground to heat water, which is more energy efficient than conventional electric water heaters.

6. **Solar water heater:** If you live in a region with ample sunlight, investing in a solar water heater can be highly beneficial. They use solar panels to capture sunlight and convert it into heat for warming water.

7. **Use a timer:** Install a timer that can turn off your electric water heater at night or when you don't use hot water.

8. **Install low-flow fixtures:** Low-flow fixtures reduce the amount of water that needs to be heated by reducing the flow rate of water through the tap, thus saving energy.

9. **Consider a hot water recirculation system:** This type of system keeps hot water circulating in your home and helps reduce wasted water while waiting for it to heat up.

In this chapter, we have reframed our understanding of the bathroom from a mere functional space to a place where we can actively commit to a sustainable lifestyle. As we continue this journey, each step takes us closer to a harmonious relationship with nature, where sustainability is not just a choice but a way of life—as vital and nurturing as the water we seek to conserve.

CHAPTER 8

Nesting Naturally: Rejecting Fast Furniture for a Sustainable Home

Do you remember Grandma's old hutch that's still holding strong after decades? It's remarkable compared to the furniture in stores today, right? This contrast is due to the rise of **"fast furniture."** Like fast fashion, furniture today is quickly and cheaply produced, focusing more on meeting the latest trends than on enduring quality. When you shop nowadays, you'll often encounter pieces that feel lightweight and transient, a far cry from the substantial heirlooms of yesteryears. This quick, disposable approach to furniture is a clear shift from our past focus on longevity and durability.

Fast furniture came as a response to a new wave of consumers who—unlike the generations before—are more mobile. These days, many of us shift from home to home (or rental to rental) and have the constant urge to update our home design choices and decor.

To put the trend into context, we buy enough furniture worldwide, equivalent to the total economic output of Sweden—$18,000 every second!

Social media platforms have helped accelerate this new trend. I always see home decor and redesign TikTok and Instagram posts where someone is redoing an area of their home simply because they were bored or needed a change.

While everyone is entitled to redo their home for whatever reason, throwing out perfectly good furniture and decor simply for a change of scenery is not sustainable for our planet!

We asked for more, cheaper furniture, and brands like IKEA, Amazon, and Wayfair were all too happy to answer the demand. What grinds my gears about all of this is that these kinds of brands count on us to keep buying cheap junky furniture from them.

We buy something cheap, it breaks, we are back to replace it within a year...and the cycle continues. Instead of offering furniture repair services, these companies know it is cheaper and easier just to buy new ones again. It's their whole business model.

With so many different furniture materials to pick from these days (some natural, some plastic), it can be confusing to dispose of properly, and most of it ends up in landfills. The EPA estimates Americans throw out over twelve million tons of furniture annually!

While the impact of a fast furniture item depends on the item itself (e.g., a couch versus a pillow), the negatives of fast furniture outweigh the positives by a landslide. From human rights abuses to clogging up our landfills, the impact of fast furniture is bad.

Planned obsolescence in fast furniture refers to the intentional design of products with a limited useful life so that consumers will need to replace them more frequently. This concept is not limited to electronics and appliances; it's also evident in the furniture industry. Here are some examples:

- **Low-quality materials:** One of the most straightforward examples is using particle board or MDF (medium-density fiberboard) instead of solid wood. These materials are more susceptible to water damage, warping, and general wear and tear. Over a short period, furniture made from these materials often becomes unstable or aesthetically unpleasing.

- **Non-replaceable components:** Some furniture pieces are designed with components, like drawer slides or hinges, that are unique to that item. If one of these components breaks or wears out, it's difficult or impossible to find a replacement, necessitating the purchase of a whole new piece of furniture.

- **Trend-driven designs:** Furniture companies may produce items that adhere strictly to fleeting fashion trends. While this doesn't represent physical obsolescence, this strategy ensures that the furniture appears "outdated" after a few years, prompting consumers to replace it to stay in style.

- **Difficult assembly and disassembly:** Some furniture is designed so that, once assembled, it's challenging or damaging to disassemble. This can be problematic for individuals who move frequently, as the furniture may not survive being taken apart and reassembled multiple times.

- **Use of glue instead of screws:** Certain furniture pieces rely heavily on adhesive for assembly rather than screws or bolts. Over time, this adhesive may degrade, making the furniture unstable.

- **Upholstery quality:** Using low-quality fabrics or thin padding that wears out, fades, or tears faster than higher-quality alternatives. Even if the frame remains sturdy, worn-out upholstery might prompt a replacement.

- **Lack of repair guides or kits:** By not providing consumers with the means or instructions to repair minor damages, companies push for the replacement of the entire piece.

- **Integrated components:** A bed frame might come with built-in lights. If those lights malfunction and they're integrated into the design, the consumer might feel compelled to replace the entire bed rather than just the light component.

The word *sustained* takes on a profound meaning in our quest for a more fulfilling life. Do we genuinely feel nourished by an endless stream of new belongings, or are we filling voids with temporary fixes? The answer may be less about acquiring and more about understanding what sustains us. Consider this: so far this year, we've produced

29,843,697 tons of PVC plastics. And the synthetic chemicals industry? It's churning out products worth nearly a billion dollars every hour, totaling the jaw-dropping value of $7.8 trillion annually—that's like combining the GDPs of Japan and the UK! With such massive figures, drilling down into specific sectors is essential, and fast furniture is a prime candidate. It might surprise you how this seemingly innocuous industry can significantly impact our planet and its people. So, let's dive in and explore the issues and what you can do about it.

1: DEFORESTATION

Wood, one of the primary materials for furniture, is under siege due to the insatiable demand driven by the fast furniture industry. The hunger for low-cost timber accelerates the degradation of the world's natural forests and exacerbates the climate and extinction crises. Deforestation, significantly fueled by this industry, is one of the top contributors to climate change. While fifteen billion trees fall annually to meet this demand, companies might argue they plant anew. Yet, their replanting often results in monoculture plantations, which barely scratch the surface of sustainability.

The environmental repercussions don't end there. Cheap wood for fast furniture is frequently sourced from tropical regions, rupturing ecosystems and eliminating trees vital to offsetting global warming. Realizing that deforestation is second only to burning fossil fuels as a climate change catalyst is alarming. Forests are crucial in climate stabilization, absorbing approximately 2.5 billion tons of CO_2 annually—nearly a third of what fossil fuels release.

Fast furniture production severely undercuts these benefits, especially with its heavy reliance on imports from China. With China's timber sourced mainly from Central Africa, the health of these forests suffers due to heightened demand. To make matters worse, a 2016 World Wildlife Fund study revealed that 68 percent of fast furniture retailers lacked timber-sourcing policies, leaving forests unprotected and consumers in the dark. Deforestation can severely harm Indigenous populations that depend on the woods for sustenance. By destroying these forests, vital habitats, food resources, and cultural legacies are lost, deeply affecting the health and survival prospects of these native communities.

2: SUPPLY CHAIN ISSUES & HUMAN RIGHTS ABUSES

Like the fast-fashion industry, the global furniture market depends on a convoluted and opaque supply chain. Such complexity, combined with a relentless demand for cheaper products, can exacerbate human rights abuses and environmental degradation, especially in the production hubs of developing nations.

One of the critical challenges in ensuring ethical furniture production is the sheer opacity of the supply chain. For instance, tracing the origins of a cotton crop used in upholsteries can be nearly impossible due to the multiple stages of production across different locations. This obscurity is similar to fast fashion, where a single garment might have components sourced from multiple countries, complicating the transparency process.

In regions like India, China, and Bangladesh, where regulations may not be as stringent, workers are often exposed to hazardous conditions reminiscent of those in the fast-fashion industry. With the pressure to produce at low costs, many face grueling hours, exposure to harmful chemicals, and limited safety protocols. This quest for affordability often comes at the price of human dignity and fair wages, leaving countless individuals trapped in exploitative working environments.

Like its fast-fashion counterpart, the furniture industry has significant environmental impacts. One of the gravest concerns is illegal logging. These illicit activities decimate forests and threaten the livelihoods of local communities that rely on these ecosystems. The scale of this issue is staggering: according to the World Wildlife Fund, illegally logged timber accounts for up to 90 percent of timber harvests in some regions, with a trade value between $51 billion and $152 billion in US currency.

3: WATER USE

Furniture production, both in scale and process, considerably impacts water resources. Cotton, frequently chosen for upholstery, is particularly water intensive, demanding as much as 2,600 gallons to yield a mere kilogram. Beyond fabric, the journey of timber, from tree to polished furniture, requires vast water quantities. This is most pronounced during the creation of wood derivatives like MDF or particle board, where the pulping process is especially water-reliant. In addition, tanneries, which produce the leather finishes beloved by many, are notorious for their water consumption and the pollution they can introduce into water systems. This water isn't just for the direct creation of furniture pieces either; it's pivotal in cooling machinery, diluting potent chemicals, and ensuring products are adequately cleaned. Consequently, the industry's water footprint expands as the global appetite for furniture grows.

4: TOXIC CHEMICALS

Fast furniture, with its lure of affordable and stylish products, comes at an environmental and health cost due to its heavy reliance on harmful chemicals. At the forefront is **formaldehyde**, commonly used in adhesives and finishes. It has garnered attention for its association with various health risks, especially when employed in rapid furniture production. Pressed wood products, particularly medium-density fiberboard (MDF), are notable culprits. The US Congress, acknowledging the gravity of the situation, passed legislation in 2010 to regulate formaldehyde emissions from such products, leading to decisive guidelines by the US Environmental Protection Agency in 2015.

Polyurethane foam, pervasive in many household items like mattresses, is environmentally and health-concerning. Sourced from petroleum, this foam releases toxins, including carcinogens like 2.4-toluene diisocyanate. Its inherent flammability necessitates treatment with chemical flame retardants. These retardants, particularly Polybrominated Diphenyl Ethers (PBDEs), have become environmental and health hazards. While some PBDEs have been phased out in the US, Europe has implemented stricter controls. Yet, in nations like China, the usage of these hazardous substances in manufacturing remains a pressing issue.

Volatile Organic Compounds (VOCs) in furniture finishes and paints pose health risks, from respiratory problems to headaches. The adverse effects of these compounds on indoor air quality have led to some regulatory scrutiny, but consistent standards are lacking globally.

PVC, a popular material in vegan leather alternatives, contains harmful components like chlorine and phthalates, known endocrine disruptors. Recognizing its potential harm, certain countries, like Canada, have initiated reviews and bans on specific phthalates, especially in children's products.

PFAS (per-and polyfluoroalkyl) chemicals, lauded for their water and stain resistance, have drawn concern due to their "forever" nature, not breaking down naturally. Recent studies have raised doubts about their effectiveness and necessity, especially in furniture fabrics. The growing evidence of PFAS's harmful effects has prompted regulatory actions, with some areas setting strict limits on their presence in drinking water. Yet, the broader challenge remains: removing PFAS from products entirely rather than managing the aftermath of their usage.

In essence, while steps have been taken legislatively to address the chemical footprint of fast furniture, a more unified and robust global response is essential to protect both human health and the environment.

5: ANIMAL RIGHTS

Animals play a significant role in sourcing and producing various furniture items. Here's a breakdown of how animals are utilized and the sustainability concerns associated with each.

Leather

Source: Leather, predominantly sourced from cows, is used for upholstery in sofas, chairs, and other furniture items.

Sustainability concerns: The livestock industry significantly contributes to greenhouse gas emissions. Tanning leather can also be environmentally damaging, especially when toxic chemicals are used. Overgrazing, associated with livestock farming, can lead to deforestation and loss of biodiversity.

Feathers and Down

Source: Feathers and down, typically from ducks and geese, are used as fillings for cushions, sofas, and bedding.

Sustainability concerns: The down extraction can sometimes be inhumane, with birds being live-plucked.

Silk

Source: Produced by silkworms, silk is sometimes used for luxurious upholstery fabrics and decorative items.

Sustainability concerns: Conventional silk production involves killing the silkworms in their cocoons. The process can be resource-intensive, requiring significant water and contributing to water pollution.

Bone and Horn

Source: Bones and horns, primarily from cattle but also from other animals, are used to make decorative inlays, handles, and knobs.

Sustainability concerns: Using animal parts can sometimes support illegal poaching, particularly when exotic animals are involved.

Wool

Source: Sheep are the primary source of wool, which is used in rugs, upholstery, and other textiles in furniture.

Sustainability concerns: Wool, though renewable, faces sustainability challenges due to overgrazing causing soil degradation and controversial practices like mulesing.

Fur

Source: Various animals, including minks and rabbits, are farmed or hunted for their fur, which might be used in specialty furniture or decorative items.

Sustainability concerns: Beyond the ethical issues of killing animals for fur, fur farming is resource-intensive and can lead to water and soil pollution.

6: WASTE

Compounding the waste issue is the complex material composition of many furniture pieces. A typical item can be made of a mix of materials, from wood and metal to various types of plastic. This diversity makes recycling a logistical nightmare. Disassembling furniture to separate and process each material is time-consuming and expensive, often rendering the effort economically unviable.

Consequently, many discarded pieces are in landfills, contributing to the growing environmental burden. Furthermore, some furniture's synthetic materials and chemical additives can leach harmful substances into the ground when disposed of, exacerbating the environmental impact. The wastage isn't limited to the end of the furniture's life; the fast-paced manufacturing processes can also lead to significant amounts of offcuts and unused materials, further contributing to the industry's waste stream.

7: POLLUTION

The fast furniture industry has significantly contributed to water and air pollution, mirroring the environmental challenges posed by its counterpart, fast fashion. For instance, the dyes and treatments used to color and finish furniture fabrics often contain harmful chemicals that, when inadequately managed, can leach into water systems. Regions heavily involved in furniture production, such as certain parts of China and Southeast Asia, have reported polluted rivers that are no longer suitable for drinking or aquatic life due to the effluents from furniture factories. Just as fast fashion has been implicated in contaminating water sources through dye run-offs, fast furniture's intensive manufacturing processes release similar pollutants. Moreover, synthetic materials, like those used in low-cost upholstery, introduce microplastics and other contaminants to water bodies, affecting marine ecosystems and entering the food chain.

Air pollution is another significant concern. Pressed wood products, commonly used in fast furniture, often involve adhesives containing formaldehyde, a volatile organic compound (VOC). When released into the air, formaldehyde contributes to smog formation and can cause human health issues. The transportation of cheap furniture products across the globe also magnifies the industry's carbon footprint, releasing greenhouse gases and exacerbating global climate change. As a contemporary example, in 2021, IKEA faced scrutiny for its logging practices in northern Russia, where clear-cutting led to reduced air quality and contributed to atmospheric carbon levels. Similarly, in Vietnam, a rising furniture production hub, the factory boom has been linked to increased levels of air and water pollutants.

HOW TO SHOP FOR SUSTAINABLE, SLOW FURNITURE

Embarking to shop for slow furniture can initially seem daunting, given the market's sea of options and claims. Yet, with the right roadmap, it becomes a rewarding pursuit of quality and sustainability. In the following sections, I'll equip you with a concise set of criteria—clear markers to guide your choices. This framework will streamline your decision-making, ensuring that every purchase aligns with the ethos of conscious consumption, durability, and timeless design.

Criteria 1: What Is It Made From?

Navigating the realm of eco-friendly furniture requires understanding the materials and fabrics that make a piece truly sustainable.

Wood has long been the go-to choice for furniture, celebrated for its enduring strength and natural beauty. The history of furniture might date back to when our ancestors first sought out alternatives to stone seats, and it's no surprise that wood became their prime choice.

In today's world, where environmental sustainability is paramount, it's crucial to recognize the ecological footprint of our choices. Hardwoods like oak, beech, and maple are often sought after for their long-lasting properties, but they take a long time to grow—sometimes hundreds of years. So, while these woods are undeniably attractive, ensuring they are responsibly sourced and as local as possible is essential. Be especially vigilant when choosing furniture crafted from rosewood, sapele, ebony, merbau, mahogany, and teak, as these materials are often harvested in an unsustainable manner. Opting for natural, unstained wood finishes is ideal, allowing you the flexibility to apply natural stains at home if desired.

On the topic of sustainability, **reclaimed wood** emerges as a strong contender. This is essentially giving old wood a new lease on life. Often sourced from retired furniture, old homes, or even unique places like submerged logs in rivers, reclaimed wood furniture not only tells a rich history but is also a testament to resource efficiency. Plus, supporting this means less new wood is harvested from old-growth forests.

Engineered wood, commonly used in furniture, is a cost-effective alternative to solid wood. The process involves compressing various wood remnants such as chips, sawdust, and shavings, then bonding them with adhesives to simulate the look and feel of genuine wood. Due to clever marketing tactics, many consumers may not immediately recognize it. Terms like MDF (medium-density fiberboard), particleboard, plywood, and chipboard often mask the true nature of the material.

One of the primary concerns with engineered wood is the adhesives involved. Many of these contain formaldehyde, which can be harmful.

However, there are variations in quality and safety among different types of engineered wood. Plywood, for instance, is composed of thin wooden layers, each oriented differently in grain direction. This arrangement offers flexibility, durability, and stability, sometimes surpassing regular solid wood.

MDF and particleboard are a different story. Derived from smaller wood fragments, including sawdust, they represent repurposed waste from other manufacturing processes. These materials don't boast the same durability as plywood, and they often require a more significant number of adhesives. This heightened adhesive use potentially introduces more formaldehyde and VOCs into the product.

It's worth noting that strides have been made to ensure safer engineered wood products. In 2010, California introduced the CARB II standards, limiting formaldehyde levels in these materials. This was followed by a nationwide standard, TSCA Title VI, effective in 2019. Consequently, engineered wood products in the US are now safer and lower in formaldehyde content.

Now, let's talk about **bamboo**. It might surprise some to learn that bamboo isn't a tree—it's grass. Yet, its rapid growth and versatility have made it a darling of sustainable designers and builders. Bamboo offers many eco-friendly possibilities with its potential to be turned into everything from floors to furniture and even entire houses. Most of it is sourced from China, where it's usually grown with minimal chemical interventions. But its swift growth can sometimes be a double-edged sword, making it essential to source responsibly. Fast harvesting can take a toll on soil health. Additionally, when considering bamboo products, it's necessary to inquire about the adhesives used, as some may contain undesirable chemicals like formaldehyde. Despite its widespread use, the green credentials of bamboo furniture are still a topic of ongoing research. Other natural materials could induce **ratan** and **wicker**, but they must be sustainably sourced.

For **upholstery**, please refer to Chapter 5 for a more detailed discussion of fabrics. Regarding furniture and home decor, look for natural fabrics like organic cotton, linen, wool, hemp, and leather.

In its many forms, leather carries a blend of advantages and drawbacks. For households bustling with kids and pets, leather stands out due to its ease of cleaning and longevity. However, when considering leather, it's essential to differentiate between chrome-tanned and vegetable-tanned varieties.

Most leather undergoes chrome tanning, a process laden with harmful chemicals like heavy metals, formaldehyde, and cyanide-based dyes. In contrast, vegetable-tanned leather avoids such toxic substances, making it the preferable choice for health-conscious consumers.

While vegetable tanning is eco-friendlier, it's also a lengthier and costlier method, making it less prevalent in large products like sofas. If a leather sofa catches your eye, it's worth inquiring about the tanning process and sourcing.

For those leaning toward ethical choices and avoiding animal products, vegan leather might seem appealing. However, one should note that most vegan leathers are essentially plastic derivatives like PU and PVC, the latter being particularly harmful. Recent tests have also revealed lead in some synthetic leather products.

Innovation has given rise to vegan leathers crafted from unique sources like pineapple, apples, and mushrooms, but furniture entirely made from such materials is still a novelty. I try to avoid synthetic fabrics like nylon, vinyl, and polyester (unless recycled) since they are made from fossil fuels, and we already know we don't need more of that in the atmosphere!

The furniture industry is experiencing a transformation driven by increasing consumer demand for sustainable options. As the spotlight turns toward eco-friendliness, many brands are diving deep into "greenovation," reimagining traditional materials and processes to pave the way for more earth-friendly choices.

Hardy Organic Hemp by O Ecotextiles, made of pest-resistant hemp, aligns with LEED criteria the US Green Building Council sets. **Knoll Textiles** offers **Abacus** resembling virgin wool woven entirely from recycled polyester, supporting eco-friendly manufacturing. **Climatex** by Rohner Textil crafts textiles from renewable sources like Ramie and beechwood, notable for their complete biodegradability and Cradle to Cradle certification. **Oliveira Textiles' Ocean Collection** blends sustainably harvested hemp with organic cotton. **Kvadrat** brings **Hallingdal**, a wool and viscose mix, abiding by stringent eco and performance criteria. Lastly, **Sensuede** utilizes recycled polyester fibers, including PET bottles, to sidestep harmful chemicals, ensuring durability and stain resistance.

Okay, so we know that **polyurethane foam** is a problematic material found in everything you are sleeping or sitting on, and finding an eco-friendly alternative is not as easy as you'd hope. Surprising? Not really! Considering it's cheap to make and cheap to buy. While polyurethane (PU) foam remains prevalent, there's an increasing shift toward more sustainable and health-conscious alternatives.

- **Natural foams:** To sidestep synthetic PU foam, consider natural foams derived from organic latex or kapok. Some eco-conscious brands have already transitioned to using these materials in their cushioning, offering consumers a more sustainable choice.

- **Soy-based foams:** Another alternative catching attention is soy foam. However, it's worth noting that soy foam, despite its "green" connotation, has its issues. The soy industry, for starters, is not without environmental controversies. Additionally, "soy-based" foams are often not purely plant-based. They might be mixed with traditional petroleum-derived foams. Despite these concerns, soy foams pose fewer environmental and health risks than standard PU foams.

- **Branded foams:** It's common for brands to market their products under enticing names like AirScape™, GelFlex™, or RestoreFlow™. However, a deeper look, often via brand FAQs or product details, reveals that these are simply variations of polyurethane foam with minor modifications.

- **Bio-foams and biodegradability:** While one might assume that plant-derived bio-foams would be biodegradable, that's not true. Surprisingly, some soy-based polyfoams might be even more resilient than their conventional counterparts, taking longer to degrade. Their unique composition makes them more challenging to recycle, especially in applications requiring uniform materials.

- **Latex foam:** The foam that genuinely stands out in the biodegradability aspect is latex foam. Made from the sap of rubber trees, it's an eco-friendly and sustainable choice.

In "slow furniture" manufacturing, utilizing recycled materials offers compelling advantages. Both **recycled metal** and **recycled fabrics** (natural and synthetic) play crucial roles in fostering sustainability. Using recycled metals such as steel and aluminum helps conserve natural resources while reducing the environmental impact of resource extraction and energy-intensive production processes. Additionally, these metals can be recycled indefinitely, contributing to a circular economy and minimizing waste. Similarly, incorporating recycled synthetic fabrics like polyester and nylon conserves valuable resources like petroleum and lowers the carbon footprint associated with their manufacturing. Moreover, by diverting discarded textiles from landfills, recycled synthetic fabrics reduce waste and alleviate environmental pollution. Using recycled plastic for furniture is highly valued, with a particular focus on HDPE sourced from plastic bottles, milk jugs, or ocean waste due to its exceptional density and durability.

Having delved deep into the world of sustainable fabrics and materials, it's time to navigate another crucial aspect of the eco-conscious journey: certifications. These emblems provide a helpful guide, but it's essential to remember they come with challenges, including being a significant financial burden for some brands. And while they serve as a helpful compass, every certification could be better. Nevertheless, as we uncover the meanings behind these labels, you'll be better equipped to make choices that align with your eco-values and the broader truths of the sustainable landscape.

FORESTRY CERTIFICATIONS

Forest Stewardship Council (FSC): Endorses sustainable forest management. It ensures products originate from forests that support biodiversity and local economies. The FSC tracks these products from forest to consumer. Their labels include: "100 percent" (from FSC forests), "Mix" (FSC and recycled materials), and "Recycled" (post-consumer and pre-consumer content).

Sustainable Forestry Initiative (SFI): A nonprofit that certifies fiber from approved forests, recycled content, or uncontested sources. It's often critiqued for having lower standards and transparency than the FSC Certification.

Programme for the Endorsement of Forest Certification (PEFC): An international nonprofit organization that promotes sustainable forest management. It operates in over thirty countries and provides certifications for forestry practices based on internationally agreed-upon standards.

Rainforest Alliance Certified: An organization that certifies products for meeting specific social, economic, and environmental sustainability standards. Their seal indicates that a product adheres to these criteria, as assessed by independent third-party auditors. They work with various products, from agricultural goods to forestry and tourism businesses.

TOXIN AND CHEMICAL EMISSIONS CERTIFICATIONS

GREENGUARD: A certification program that identifies products with low chemical emissions, ensuring better indoor air quality. Covering items like building materials, furniture, electronics, and specific medical devices, this certification helps consumers choose products that contribute less to indoor air pollution due to reduced VOC emissions.

GREENGUARD Gold is an enhanced certification that sets stricter criteria for VOC emissions, ensuring products meet health-based standards suitable for sensitive environments like schools and healthcare facilities. Beyond the restrictions on over 360 VOCs and overall chemical emissions, products with this certification also adhere to the requirements set by California's Department of Public Health.

The **Formaldehyde Free Verified** certification is awarded by UL Environment, the company responsible for GREENGUARD labels. This certification focuses on determining if a product is entirely free from formaldehyde.

MADE SAFE: This certification ensures that products, from apparel and bedding to personal care and childcare items, are devoid of over 6,500 harmful chemicals. To earn this certification, companies provide a comprehensive ingredient list for evaluation. Initially, the list is cross-referenced against banned chemicals. Subsequent assessments involve a toxicant database screening to authenticate certain chemicals. Before a final report is issued, further evaluations encompass aspects like bioaccumulation, environmental impact, and potential human health effects.

CertiPUR-US: An industry certification designed for polyfoam products to ensure they're made with fewer harmful chemicals. While it does confirm the foam is free from specific flame retardants like PDBE and has reduced levels of formaldehyde, phthalates, and certain heavy metals, the certification has been criticized by some as a "greenwashing" tactic. Although a CertiPUR-US certified foam might improve over standard memory foam, it's still predominantly synthetic, resource-intensive, and can emit VOCs. Consumers should be cautious and consider the broader context when evaluating a product's safety and environmental implications.

OEKO-TEX certifies textiles and leather for safety and sustainability. Their Standard 100 ensures every part of a product, down to threads and buttons, is tested for harmful substances. The leather standard focuses on toxic chemicals, specifically in leather. The Made in Green label, applicable to all textiles, ensures products meet chemical standards and are produced sustainably and ethically. Each Made in Green item also offers traceability for consumer transparency.

Eco-INSTITUT: A German certification that focuses on products affecting indoor air quality due to off-gassing. They award their label to low-pollutant and low-emission products. Testing involves examining emissions and analyzing for heavy metals, pesticides, and other specific compounds, varying based on the product type.

MATERIAL AND PRODUCTION PROCESS CERTIFICATIONS

Cradle to Cradle Certified: a sustainability label for products designed for the circular economy. Founded on principles from the book *Cradle to Cradle: Remaking the Ways We Make Things*, it evaluates products in five areas: material health, material reuse, renewable energy and carbon management, water stewardship, and social fairness. Products receive ratings from Basic to Platinum in each category, with details publicly accessible on the organization's website.

Global Organic Latex Standard (GOLS): Certifies that latex products like mattresses and pillows contain over 95 percent certified organic raw latex. It ensures the final product passes emission tests and restricts certain harmful substances. The standard encompasses the entire supply chain, from rubber plantations to retailers, monitored through transaction certificates.

Global Organic Textile Standard (GOTS): Certifies textile products based on their organic content and ethical production process. There are two levels: "organic" for products with 95 percent or more organic material and a second for those with 70 to 94 percent organic content. The certification covers the entire supply chain, ensuring transparency and traceability. Each GOTS label includes the license number or supplier name, allowing consumers to verify product details.

Bluesign®: A certification for sustainable textiles, ensuring that the materials and the production process meet high environmental and safety standards. To bear the label, a product must have at least 90 percent Bluesign-approved textiles and 30 percent approved accessories. The certification assesses the entire textile supply chain to guarantee sustainability from raw material to the finished product.

Sustainable Furnishings Council: A membership organization that promotes sustainable practices in the furnishings industry. While not a certification, member companies must adhere to specific environmental and production standards. Unlike the Forest Stewardship Council, which focuses on wood, the SFC evaluates a company's environmental impact. Though certifications are valuable indicators, lacking them doesn't automatically imply a brand uses harmful materials.

USDA Organic: Indicates that the material, like cotton or hemp, has been grown without synthetic fertilizers, pesticides, and genetically modified organisms (GMOs). For furniture, this often pertains to organic upholstery or filling materials.

RECYCLING AND END-OF-LIFE MANAGEMENT

The Recycled Claim Standard (RCS) and the Global Recycled Standard (GRS) are guidelines for validating recycled content in products. Both aim to standardize "recycled" definitions, verify recycled content, and inform brand and consumer choices. The GRS is more rigorous, demanding a minimum of 50 percent recycled content and encompassing broader environmental and social criteria. The Textile Exchange oversees these standards.

Animal Welfare Certifications include the Responsible Wool Standard (RWS), Leather Working Group (LWG), and the Responsible Down Standard (RDS). Refer to Chapter 5 for an overview of these standards.

Criteria 2: Sustainable Sourcing/ Ethical Manufacturing

That chair you are sitting on or the bed you are sleeping on should not only provide comfort but also stand as a testament to ethical sourcing. Every piece of furniture in our homes carries a story, from the trees cut to make them to the hands that crafted them and the methods used in their transportation. Ethical sourcing ensures that this story is one of respect: respect for the environment, for the rights of workers, for local communities, and for future generations.

ETHICAL LABOR & FAIR TRADE CERTIFICATIONS

Fair Trade Certified is a global brand linked to Fair Trade USA, a nonprofit. There's an agreement with Fairtrade International to prevent repeated audits. Producers with this seal may have passed standards set by either Fair Trade USA or Fairtrade International. This certification can apply to an entire product, a component, or the production facility. It stands for income sustainability, community well-being, empowerment, environmental stewardship, workers' rights, and sustainable land management.

Fairtrade International manages the Fairtrade mark, distinct from Fair Trade Certified. It endorses over 30,000 products and has a combined standard focused on supporting the sustainable development of producers in the Global South. It emphasizes environment-friendly practices, bans certain pesticides and GMOs, advocates equitable profit distribution, and ensures proper working conditions while prohibiting child labor and discriminatory practices.

Ethical Trading Initiative (ETI): A UK-established independent organization formed in 1998. It unites companies, trade unions, and NGOs to promote international labor standards within the global supply chains of its members. All members are mandated to uphold the ETI base code, setting the foundation for workers' rights and conditions.

The Fair Rubber Association oversees the Fair Rubber certification, ensuring natural latex products meet Fair Trade criteria. This certification guarantees that rubber is procured to support better living and working conditions for rubber farmers and promotes environmentally friendly, chemical-free rubber production.

International Labor Organization (ILO): A UN agency that promotes social and economic justice through international labor standards. These standards, outlined in 189 conventions and treaties, focus on ensuring fair, secure, and dignified work conditions globally. Eight are deemed fundamental, safeguarding rights such as freedom of association, collective bargaining, and preventing forced labor, child labor, and workplace discrimination. The ILO plays a pivotal role in shaping international labor law.

So, given all that, have you ever considered taking your commitment to ethical sourcing a step closer to home? I've been thinking a lot about the power of buying locally, especially regarding our furniture. Here's the scoop: local pieces often have a unique charm that's hard to find in big chain stores. And the craftsmanship? Top-notch. Local artisans aren't just making furniture; they create stories, often using superior materials and techniques.

And get this: shopping local is like giving your community a warm, friendly hug. You're directly supporting the people who might be your neighbors or friends. It's heartwarming to think our purchases can help local families thrive. Plus, nature seems to be happier too! Many local makers prioritize eco-friendly practices, so we're doing our part for the planet.

One more thing: ever heard of a "local discount"? Some artisans offer fantastic deals, so you're getting brilliant quality at a steal.

Criteria 3: Corporate Social Responsibility

In furniture design and manufacturing, corporate responsibility is about more than just turning a profit—it's about creating products that stand the test of time, both in style and durability, while minimizing harm to our planet. Companies that are genuinely committed to a green future go beyond the basics. They think about the lifecycle of their products, from sourcing raw materials to what happens to a piece of furniture when it's no longer needed. A hallmark of such commitment is seen in brands that adhere to the Cradle 2 Cradle (C2C) certification. These companies ensure their products can be disassembled and recycled, paving a path toward a circular economy where waste becomes a thing of the past.

Moreover, durability is a cornerstone of corporate responsibility. Brands that invest in crafting furniture that lasts are making an eco-friendly choice and showing a commitment to consumer value. A table that has stood firm for decades tells a story of quality materials, skilled craftsmanship, and a brand ethos centered around longevity, not disposability. Furthermore, forward-thinking brands understand the evolving needs of modern consumers. Living spaces are transforming, and the demand for flexible and space-saving solutions is growing. As part of their responsibility, companies are innovating to provide lightweight, foldable, and multifunctional furniture options. And by supporting handcrafted or custom-made pieces, they contribute to a reduced carbon footprint and a more bespoke consumer experience. Packaging is an often overlooked yet critical aspect of sustainability. Enlightened companies are adopting packaging methods that use fewer resources, are biodegradable, and minimize transportation emissions. Addressing packaging waste means considering every step of a product's journey, from extraction to disposal. As the conversation on environmental responsibility evolves, many companies explore carbon offsetting and claiming carbon neutrality. While these initiatives can signal a company's commitment to reducing its carbon footprint, they have pros and cons.

On the one hand, carbon offsets can support essential environmental projects and innovations. On the other hand, they can sometimes be used to sidestep genuinely sustainable changes in operations. Consumers and stakeholders should scrutinize these claims and understand their full implications to ensure they're not just greenwashing tactics.

CERTIFICATIONS

B Corp Certification, or Certified B-Corporation, is a title awarded to companies that meet the highest social and environmental performance standards, transparency, and accountability. Currently, over 3,800 companies across seventy-four countries have achieved this certification. These companies undergo a comprehensive B Impact Assessment, which evaluates their impact on the environment, workers, community, and customers. Once certified, their performance metrics are publicized on the B-Corporation website. Additionally, B Corp-certified companies must amend their governing documents to align their board of directors' mission with B Corp's ethos of balancing profit with purpose.

1 percent for the Planet: Founded by Yvon Chouinard of Patagonia and Craig Mathews of Blue Ribbon Flies, 1 percent for the Planet encourages companies to donate 1 percent of their gross sales to environmental charities. While the initiative has generated over $250 million for environmental causes, participating doesn't guarantee a company's products are all eco-friendly. The organization also promotes individual salary donations.

Green America's Green Business Certification: Recognizes businesses prioritizing social and environmental welfare. It isn't strictly about products but rather the company's overarching ethos. Certified firms must use their platform for positive societal change, adhere to eco-friendly practices, ensure social equity, and operate with utmost transparency. The certification covers many businesses, from banks and resorts to candlemakers, requiring each to meet specific standards tailored to their industry.

Leadership in Energy and Environmental Design (LEED): Managed by the US Green Building Council, assesses and certifies eco-friendly construction practices. Projects earn points based on environmental strategies and can be classified as LEED Certified, Silver, Gold, or Platinum. LEED Zero is the highest standard for net zero carbon or resource projects. The program also evaluates cities' environmental metrics, including water use and transportation.

SUSTAINABLE FURNITURE CHECKLIST

- ☐ Natural Fabrics & Materials
- ☐ Certifications
- ☐ Low or No Formaldehyde
- ☐ No Flame Retardants
- ☐ Low or No VOC Finishes
- ☐ No PFAS
- ☐ Recycled Components
- ☐ Made/Sourced Locally

BRAND RECOMMENDATIONS

Several brands are leading the pack when considering sustainable, "slow furniture" options.

Sabai is recognized for avoiding harmful chemicals and promoting furniture longevity with initiatives like their Repair and Revive programs, all while sourcing 90 percent of their materials locally. Based in Los Angeles, **Medley** champions eco-friendly materials, including FSC-certified wood and natural latex; they also provide consumer-friendly financing options. Avocado has many third-party certifications and a commitment to zero-waste, carbon-negative designs.

Ecobalanza caters to a range of consumer preferences, from those with allergies to vegans, and their handcrafted designs come with several eco certifications.

Natural Home by The Futon Shop shuns petroleum-based products in favor of natural alternatives, offering a range of designs and holding notable environmental certifications. These brands exemplify responsible, sustainable furniture production in today's market.

The Citizenry emphasizes materials like organic cotton, alpaca wool, and FSC-certified wood, with items made through fair trade practices across the globe, as certified by the World Fair Trade Organization.

Ten Thousand Villages brings a wide array of eco-friendly decor from recycled saris, hex nuts, and sustainable Kisii soapstone. As a founding member of the WFTO, they've been pivotal in advancing fair trade principles worldwide, investing over $100 million toward empowering artisans in developing countries.

Parachute stands out with its selection of natural materials such as cotton, linen, and organic jute, many of which bear the OEKO-TEX certification. Their global manufacturing footprint spans family-owned factories in Europe and the US Emphasizing environmental stewardship, they're Climate Neutral–certified and employ recyclable shipping practices.

Ocelot Market is renowned for its vast array of sustainable fabrics and materials, including 100 percent recycled chenille scraps. Their commitment to ethical labor practices is evident in their support for fair trade goods and partners with transparent supply chains.

Lastly, **VivaTerra** showcases many sustainable materials, from ceramic clay to reclaimed wood. They uphold ethical criteria and fair trade practices. Their green initiatives include a "climate-friendly shipping" option, allowing customers to offset their carbon footprint easily.

When evaluating the safety of **IKEA** furniture, it's essential to balance its benefits against potential drawbacks. On the positive side, IKEA refrains from using PFAS, a group of harmful chemicals, and only employs flame retardants where legally mandated, not including the US, as per California's 2020 guidelines. Based in Sweden, the company adheres to the stringent European Union chemical safety standards, arguably among the most rigorous globally. Furthermore, IKEA maintains a comprehensive list of restricted substances, setting it apart from many mainstream brands, all while offering affordability. However, there are concerns. IKEA frequently employs engineered wood, known for elevated formaldehyde levels. The brand's transparency on ingredients in paints and finishes sometimes leaves consumers wanting.

Additionally, they often resort to synthetic, petroleum-derived materials. The safety level can vary, depending on the specific product, necessitating informed decision-making by buyers. Third-party certifications are absent in IKEA's portfolio, which might raise eyebrows for those seeking external validation. In sum, while IKEA has made commendable strides in offering safer furniture, consumers should approach it with awareness and, if necessary, a little extra research.

When sprucing up our homes, it's tempting to splurge. But here's a secret: furnishing on a budget doesn't mean compromising style or quality. Start with a "less is more" mindset. Ask yourself, "Do I truly need this?" before adding another piece to your collection. Embracing minimalism saves cash and makes your rooms appear larger and more elegant. And if you've got an eagle eye, you might find a diamond in the rough—those items with slight imperfections, often sold at discounts, that only need a bit of love. Please don't shy away from seasonal sales or outlets; they're goldmines for fantastic finds at a fraction of the price. Got an old chair or table? A revamp might be all it needs. With some paint or new fabric, it could become the centerpiece of your room. Remember the charm of bartering? It's alive and well—swap items you don't need anymore with ones you've been coveting. And while bargains are always a thrill, prioritize quality over quick deals. Investing in sturdy items that'll stand the test of time is smarter than replacing flimsy pieces. Lastly, if space is a concern, pick furniture that does double duty—think beds with drawers or ottomans that double as storage. In short, with a blend of creativity, patience, and intelligent strategies, you can achieve that magazine-worthy home without emptying your wallet.

Arguably, the most environmentally friendly pieces are those which already exist. Reusing and repurposing furniture reduces the demand for new resources and minimizes waste destined for landfills, but there are a few things to keep in mind.

Once you decide to venture into vintage furniture, setting a clear budget is essential. Remember, used furniture purchases only sometimes come with the financing options that traditional retailers might offer. Always do your due diligence by checking sites like SaferProducts.gov to ensure the item hasn't been recalled. If it was, demand proof from the seller that the necessary repairs have been made.

It's a mantra worth repeating: always pay attention to the details. Carefully inspect each item that catches your eye. Is the chair missing a screw? Does that table have an unmentioned scratch? It's vital to determine if these flaws are something you can rectify or if you can live with them. Naturally, you'll want items that promise longevity and durability. And while it might sound basic, always measure your intended space and cross-reference those measurements with the item's dimensions to avoid any fitment issues.

Secondhand or vintage furniture shopping requires patience. You may not strike gold with the first listing you see. Equally, be on your guard for scams, especially on platforms like Facebook Marketplace.

On the topic of older furniture, there's a prevalent notion that buying older items eliminates concerns about outgassing or off-gassing chemicals. This is only partially accurate. Even if once treated with chemicals, older solid wood furniture is less likely to compromise air quality. However, they might need freshening up due to accumulated odors. Conversely, older upholstered items like sofas can still release harmful toxins from the foam, glues, and fabric treatments.

Moreover, older furnishings might even harbor molds. It's interesting to note that while less furniture today is made with flame-retardant chemicals, older pieces might still be culprits. Some estimates suggest that up to 85 percent of US couches bought between 1984 and 2010 contain these potentially harmful chemicals. These toxins can emanate from the foam, proving toxic throughout the furniture's lifespan. There's also a risk associated with vintage or salvaged furniture: bed bugs. Hence, it's advisable to be cautious when buying old couches or mattresses. Instead, explore reliable platforms such as AptDeco, Kaiyo, Etsy's reclaimed section, and Chairish for safer secondhand furniture options.

In the beginning chapters of this book, we journeyed into the heart of the circular economy—a blueprint for a world where resources are used, recycled, and reused with minimal waste. The "slow furniture" philosophy beckons with renewed urgency as we step into home decor and furnishings. Much like its counterpart in the fashion world, slow furniture is more than just a trend; it's a call to action, a conscious choice to champion the pillars of sustainability, durability, and mindful consumption. By choosing furniture designed to last and be cherished, we're not just furnishing our spaces; we're anchoring our lives with pieces that tell stories of craftsmanship and conscious choices. This idea of being *sustained* goes beyond the mere longevity of an item; it speaks of a holistic approach where our well-being is interwoven with that of the environment and the artisans behind each creation. Amidst the cascade of challenges birthed by fast furniture's relentless pace, the commitment to slow furniture stands as a poignant reminder that the solutions to our overconsumption dilemmas may lie in slowing down, appreciating more, and consuming less.

CONCLUSION

Tipping Point: The Urgent Need for Sustainable Choices

Throughout this book, I've endeavored to equip you with information to make informed, "mindful" choices. However, the crux of living a *sustained* life is that we need to purchase less. Being mindful is the cornerstone of sustainability. When choosing an item, we must remember that someone invested time and effort into creating it. We also need to recognize that there's no "away." Every item we discard goes somewhere, often with environmental repercussions.

Living sustainably often comes with a price tag, a reality many are all too familiar with. While more sustainable and ethically made products generally cost more, it's essential to remember that this isn't an invitation to rush out and make new purchases, even if they are sustainable. Systemic issues, such as growing wealth disparities, contribute to this pricing dynamic. While I advocate for sustainable choices throughout this book, it's not my intention to induce guilt or shame for anyone who finds these options out of their financial reach. The vision is for sustainable living to become a universally accessible norm. Until that ideal is realized, the most genuine advice is to do your best with the resources you have at hand. Remember, the journey to sustainability doesn't demand perfection; it values earnest effort.

Climate change is a complex, multidimensional issue whose main drivers include electricity generation, manufacturing, deforestation, transportation, food production, building operations, and consumer behavior. Most of our electricity is still produced by burning carbon dioxide and nitrous-oxide-emitting fossil fuels, with only about a quarter coming from cleaner, renewable sources.

Similarly, industrial processes, particularly those involved in manufacturing goods like cement, steel, electronics, and plastics, rely heavily on burning fossil fuels for energy, making this sector a significant contributor to global emissions. These industries, including mining and construction, release harmful gases during their operation. The rampant cutting down of forests exacerbates this problem. Forests, which act as natural carbon sinks, release stored carbon when cut while reducing nature's ability to absorb emissions.

The transportation sector is another significant source of greenhouse gases. Most vehicles still run on fossil fuels, resulting in carbon dioxide emissions. This problem extends to sea and air transport as well. Food production is another major contributor to climate change due to deforestation for agriculture, methane produced by ruminant digestion, energy use in farming, and emissions from food packaging and distribution.

In addition, the energy consumed by residential and commercial buildings, primarily for heating, cooling, and powering appliances, is another major contributor. This is particularly troubling given the increase in energy consumption due to the rise in air-conditioner ownership and electricity consumption for various appliances. Moreover, individual consumption patterns significantly contribute to greenhouse gas emissions, with the wealthiest 1 percent of the global population responsible for more emissions than the poorest 50 percent.

The effects of climate change are multifaceted and severe. Rising greenhouse gas concentrations have led to higher global temperatures, more intense storms, increased drought, and a warming, rising ocean. This has resulted in species loss, food shortages, health risks, and increased poverty and displacement. As the earth continues to warm, extreme weather events are becoming more common, disrupting ecosystems, diminishing food supplies, and causing massive displacement of people. Climate change is humanity's most significant health threat, with changing weather patterns exacerbating diseases and extreme weather events, making it difficult for healthcare systems to cope.

Our industrial and individual behaviors are contributing to climate change, and the impact is already visible in the form of increased temperatures, extreme weather events, species loss, and rising poverty. Without a doubt, urgent action is needed on multiple fronts to tackle this existential threat.

We are at a tipping point. Climate scientists have recently issued a "code red for humanity," highlighting the dire need to curb global warming below 1.5°C. Yet many nations, including some in the G7, have increased their emissions since the Paris Agreement. We must urgently urge our governments to take monumental steps in addressing climate change. The time for passive diplomacy has passed. Indigenous communities are pivotal in climate solutions with their profound ecological understanding and intrinsic connection to nature. These communities have historically faced the brunt of environmental injustices and racism, yet they've been at the forefront, protecting lands against exploitative activities. We can be allies in the truest sense by supporting organizations like Indigenous Climate Action and land defenders like the Wet'suwet'en. Moreover, it's vital to educate oneself about truth and reconciliation and integrate anti-colonial perspectives into our lives and advocacy.

The clean-energy shift is pivotal. Whether you can afford to install solar panels, you can still participate in this revolution. Local renewable energy co-ops offer opportunities to invest in cleaner futures. Moreover, communicate with financial advisers to ensure your investments are not tied to fossil fuels. If institutions you're associated with support fossil fuels, it's time to rally for divestment. **While individual actions matter, systemic change is paramount.** It's our duty to elect leaders who prioritize climate change mitigation. They must advocate for science-backed targets, adaptations, and transitions to clean energy. Every vote counts, so register, research, and participate actively in elections.

If you're underage, encourage educational institutions to participate in programs that give a taste of the democratic process.

Coined by Leah Thomas, intersectional environmentalism stresses the intertwined nature of social and environmental injustices. It acknowledges that marginalized communities often bear the brunt of environmental degradation. This environmentalism champions justice for both people and the planet. Environmental racism exposes communities of color and low-income groups to disproportionate environmental hazards. This is not just an environmental issue but one deeply embedded in systemic racism. It's essential to understand this in the broader context of colonialism, where societal structures have historically uplifted certain groups at the expense of others.

In essence, we need to act collectively and swiftly. The planet is not just warming; it's on fire. Complacency is not an option anymore. Let us commit to a *sustained*, balanced, and just approach to our economic, social, and environmental responsibilities. Let's pledge to leave behind a world where future generations can survive, thrive, and sustain themselves.

Acknowledgments

A special thank you to my husband Bodhi, who taught me that every summit is reached by taking the first step. I could never have done any of this without you by my side.

To my mom, the thrifting queen who showed me the joy of discovering hidden treasures, and to my stepdad Joe, who instilled in me the value of recycling and finding worth in what we keep.

To my dad, who taught me the invaluable lesson that prior planning prevents poor performance.

To Mango for helping me throughout my writing journey.

And finally, a special thanks to my readers. Without you, this book would be like a library without any visitors—full of stories but no one to tell them to!

About the Author

Meet **Candice Batista**. Award-winning environmental journalist. Revered eco-living authority.

Having devoted her entire career to environmental journalism, Candice knows a thing or three about how our daily choices impact wildlife, the climate, and the environment. Candice has spent years on the front lines of national and international media as one of Canada's leading eco advocates, leveraging her background in media and communications to report on the most pressing climate and environmental crises for digital audiences worldwide.

Batista, hailed as the leading eco expert in Canada for over twenty-five years, has made several television appearances on platforms including HuffPost, *The Globe and Mail*, The Weather Network, and The Pet Network. She has also been featured on CTV's *The Marilyn Denis Show* and *Your Morning*, and she is currently the leading sustainability expert for City TV's *Cityline* and *Breakfast Television*.

Candice is also the proud founder and owner of The Eco Hub, a leading online destination and resource for all things sustainable. The Eco Hub has spent close to a decade as a well-regarded lifestyle and news publication with a growing audience of green living advocates.

Candice has also produced over five thousand TV segments and three television productions centered around sustainable living, including *A Greener Toronto*, *@issue Earth*, and *Global Footprints*. In 2011, the City of Toronto nominated Batista for their Green Toronto Awards in the Environmental Awareness category.

Batista is also one of a few Canadians to be trained by former US Vice President Al Gore to deliver the powerful Inconvenient Truth presentation. She is also a member of the "Fashion Takes Action" advisory board and a proud spokesperson for World Animal Protection.

Bibliography

"68% of the World Population Projected to Live in Urban Areas by 2050, Says UN | UN DESA | United Nations Department of Economic and Social Affairs," n.d. www.un.org/development/desa/en/news/population/2018-revision-of-world-urbanization-prospects.html.

"A New Textiles Economy: Redesigning Fashion's Future," n.d. ellenmacarthurfoundation.org/a-new-textiles-economy.

Admin. "Why Isn't the U.S. Recycling Rate Growing | CompuCycle." *CompuCycle* - (blog), December 3, 2021. compucycle.com/blog/why-isnt-the-u-s-recycling-rate-growing/#:~:text=12.3%20million%20tons%20of%20glass,million%20tons%20went%20to%20landfills.

US EPA. "Advancing Sustainable Materials Management: Facts and Figures Report | US EPA," April 4, 2023. www.epa.gov/facts-and-figures-about-materials-waste-and-recycling/advancing-sustainable-materials-management.

Green Science Policy. "Analysis: Most Research on PFAS Harms Is Unpublicized - Green Science Policy Institute," n.d. greensciencepolicy.org/news-events/press-releases/analysis-most-research-on-pfas-harms-is-unpublicized.

Barrett, Julia R. "Chemical Exposures: The Ugly Side of Beauty Products." *Environmental Health Perspectives* 113, no. 1 (January 1, 2005). doi.org/10.1289/ehp.113-a24.

Beall, Abigail. "Why Clothes Are so Hard to Recycle." *BBC Future*, February 24, 2022. www.bbc.com/future/article/20200710-why-clothes-are-so-hard-to-recycle.

Betts, Kellyn S. "New Thinking on Flame Retardants." *Environmental Health Perspectives* 116, no. 5 (May 1, 2008). doi.org/10.1289/ehp.116-a210.

The Future Market. "Biodiversity — The Future Market," n.d. thefuturemarket.com/biodiversity#:~:text=About%2075%25%20of%20 the%20world's,%2C%20pests%2C%20and%20climate%20change.

Browne, Mark a. Oakley, Phillip Crump, S. J. Niven, Emma L. Teuten, Andrew Tonkin, Tamara S. Galloway, and Richard C. Thompson. "Accumulation of Microplastic on Shorelines Worldwide: Sources and Sinks." *Environmental Science & Technology* 45, no. 21 (October 4, 2011): 9175–79. doi.org/10.1021/es201811s.

Center for Devices and Radiological Health. "Menstrual Tampons and Pads: Information for Premarket Notification Submissions (510(k)s) - Guidance for Industry and FDA Staff." *U.S. Food And Drug Administration*, June 29, 2018. www.fda.gov/regulatory-information/search-fda-guidance-documents/menstrual-tampons-and-pads-information-premarket-notification-submissions-510ks-guidance-industry.

"CFDA," n.d. cfda.com/resources/materials.

Greenpeace USA. "Circular Claims Fall Flat Again - Greenpeace USA," October 26, 2022. www.greenpeace.org/usa/reports/circular-claims-fall-flat-again.

Cline, Elizabeth. "Where Does Discarded Clothing Go?" *The Atlantic*, July 18, 2014. www.theatlantic.com/business/archive/2014/07/where-does-discarded-clothing-go/374613.

World Wildlife Fund. "Cotton | Industries | WWF," n.d. www.worldwildlife.org/industries/cotton.

WWF. "Deforestation," n.d. wwf.panda.org/discover/our_focus/forests_practice/deforestation.

Dmizen. "The July 2015 Timber Scorecard." *WWF*, November 10, 2016. www.wwf.org.uk/node/32561.

US EPA. "Durable Goods: Product-Specific Data | US EPA," December 3, 2022. www.epa.gov/facts-and-figures-about-materials-waste-and-recycling/durable-goods-product-specific-data#:~:text=To%20measure%20the%20generation%20of,2.2%20million%20tons%20in%201960.

FPS Public Health. "Effect of Detergents on the Environment," May 17, 2021. www.health.belgium.be/en/effect-detergents-environment.

FPS Public Health. "Effect of Detergents on the Environment," May 17, 2021. www.health.belgium.be/en/effect-detergents-environment.

European Food Safety Authority. "Brominated Flame Retardants." *European Food Safety Authority*, June 8, 2023. www.efsa.europa.eu/en/topics/topic/brominated-flame-retardants#:~:text=EU%20framework&text=Directive%202003%2F11%20%2FEC%2C,higher%20than%200.1%25%20by%20mass.

"Fabric Softeners & Conditioners Market Size, Share & Trends Analysis Report By Product (Liquid, Dryer Sheets), By Application (Household, Commercial), By Region, And Segment Forecasts, 2019 - 2025," n.d. www.grandviewresearch.com/industry-analysis/fabric-softeners-conditioners-market.

UN News. "FEATURE: UN's Mission to Keep Plastics out of Oceans and Marine Life," October 11, 2021. news.un.org/en/story/2017/04/556132-feature-uns-mission-keep-plastics-out-oceans-and-marine-life.

Federation, British Plastics. "Polyvinyl Alcohol (PVOH)." British Plastics Federation, n.d. www.bpf.co.uk/plastipedia/polymers/polyvinyl-alcohol-pvoh.aspx.

IUCN. "Forests and Climate Change," n.d. www.iucn.org/resources/issues-brief/forests-and-climate-change.

National Cancer Institute. "Formaldehyde - Cancer-Causing Substances," December 5, 2022. www.cancer.gov/about-cancer/causes-prevention/risk/substances/formaldehyde.

US EPA. "Formaldehyde Emission Standards for Composite Wood Products | US EPA," February 21, 2023. www.epa.gov/formaldehyde/formaldehyde-emission-standards-composite-wood-products.

Fourcassier, Sarah, Mélanie Douziech, Paula Perez-Lopez, and Londa Schiebinger. "Menstrual Products: A Comparable Life Cycle Assessment." *Cleaner Environmental Systems* 7 (December 1, 2022): 100096. doi.org/10.1016/j.cesys.2022.100096.

US EPA. "Frequent Questions about Landfill Gas | US EPA," April 21, 2023. www.epa.gov/lmop/frequent-questions-about-landfill-gas.

Greenpeace International. "Black Friday: Greenpeace Calls Timeout for Fast Fashion - Greenpeace International," November 24, 2016. www.greenpeace.org/international/press-release/7566/black-friday-greenpeace-calls-timeout-for-fast-fashion.

www.fmi.org. "Grocery Industry Launches New Initiative to Reduce Consumer Confusion on Product Date Labels," February 15, 2017. www.fmi.org/newsroom/news-archive/view/2017/02/15/grocery-industry-launches-new-initiative-to-reduce-consumer-confusion-on-product-date-labels.

Hoskins, Tansy. "Cotton Production Linked to Images of the Dried up Aral Sea Basin." *The Guardian*, August 25, 2021. www.theguardian.com/sustainable-business/sustainable-fashion-blog/2014/oct/01/cotton-production-linked-to-images-of-the-dried-up-aral-sea-basin.

Recycle Now. "How Is Glass Recycled? | Recycle Now," n.d. www.recyclenow.com/how-to-recycle/glass-recycling.

Hub, IISD's Sdg Knowledge. "World Bank Report Warns Global Solid Waste Could Increase 70 Percent by 2050 | News | SDG Knowledge Hub | IISD," n.d. sdg.iisd.org/news/world-bank-report-warns-global-solid-waste-could-increase-70-percent-by-2050/#:~:text=The%20report%20titled%2C%20'What%20a,urbanize%20and%20increase%20their%20populations.

Kart, Jeff. "Study Says Up to 75% Of Plastics From Detergent Pods Enter The Environment, Industry Says They Safely Biodegrade." *Forbes*, August 8, 2021. www.forbes.com/sites/jeffkart/2021/08/08/study-says-up-to-75-of-plastics-from-detergent-pods-enter-the-environment-industry-says-they-safely-biodegrade/?sh=57726b40796a.

Kavilanz, Parija. "After Plastic Straws, This Entrepreneur Wants Plastic Toothbrushes to Disappear." CNNMoney, July 21, 2018. money.cnn.com/2018/07/21/news/economy/bamboo-toothbrush-plastic-alternative#:~:text=According%20to%20advocacy%20group%20Plastic,plastic%20toothbrushes%20in%20that%20time.

Kaza, Silpa, Lisa Yao, Perinaz Bhada-Tata, and Frank Van Woerden. *What a Waste 2.0: A Global Snapshot of Solid Waste Management to 2050*. Washington, DC: World Bank EBooks, 2018. doi.org/10.1596/978-1-4648-1329-0.

Kilgore, Georgette. "How Many Trees Cut down Each Year or in 2023? The Deforestation Crisis Explained." *8 Billion Trees: Carbon Offset Projects & Ecological Footprint Calculators*, July 10, 2023. 8billiontrees.com/trees/how-many-trees-cut-down-each-year.

Kim, Min Joo, and Young Joo Park. "Bisphenols and Thyroid Hormone." *Endocrinology and Metabolism* 34, no. 4 (January 1, 2019): 340. doi.org/10.3803/enm.2019.34.4.340.

Knox, Kristin, Robin E. Dodson, Ruthann A. Rudel, Claudia Polsky, and Megan R. Schwarzman. "Identifying Toxic Consumer Products: A Novel Data Set Reveals Air Emissions of Potent Carcinogens, Reproductive Toxicants, and Developmental Toxicants." *Environmental Science & Technology* 57, no. 19 (May 2, 2023): 7454–65. doi.org/10.1021/acs.est.2c07247.

AP News. "Laundry Detergent Market Size Analysis, Trends, Top Manufacturers, Share, Growth, Statistics and Forecast to 2025 | AP News," December 18, 2021. apnews.com/press-release/

wired-release/business-wyoming-sheridan-clorox-co-unilever-plc-aa15c035e533c56ba9a6d60c3585060c.

Legislative Services Branch. "Consolidated Federal Laws of Canada, Phthalates Regulations," June 22, 2016. laws-lois.justice.gc.ca/eng/regulations/SOR-2016-188/page-1.html#:~:text=3%20(1)%20The%20vinyl%20in,DNOP)%20when%20tested%20in%20accordance.

Levy, Stuart B. "Antibacterial Household Products: Cause for Concern." *Emerging Infectious Diseases* 7, no. 7 (June 1, 2001): 512–15. doi.org/10.3201/eid0707.017705.

MacInnis, Gail, Étienne Normandin, and Carly Ziter. "Decline in Wild Bee Species Richness Associated with Honey Bee (*Apis Mellifera* L.) Abundance in an Urban Ecosystem." *PeerJ* 11 (February 3, 2023): e14699. doi.org/10.7717/peerj.14699.

McGivney, Annette. "'Like Sending Bees to War': The Deadly Truth behind Your Almond Milk Obsession." *The Guardian*, October 29, 2021. www.theguardian.com/environment/2020/jan/07/honeybees-deaths-almonds-hives-aoe.

Environmental Working Group. "'Meets All Government Standards': EWG'S 2019 Tap Water Database Details Unsafe Contamination in Communities Nationwide," October 23, 2019. www.ewg.org/news-insights/news-release/meets-all-government-standards-ewgs-2019-tap-water-database-details.

US EPA. "National Overview: Facts and Figures on Materials, Wastes and Recycling | US EPA," December 3, 2022. www.epa.gov/facts-and-figures-about-materials-waste-and-recycling/national-overview-facts-and-figures-materials.

Nutrition, Center for Food Safety and Applied. "Food Loss and Waste." *U.S. Food And Drug Administration*, February 14, 2023. www.fda.gov/food/consumers/food-loss-and-waste#:~:text=In%20the%20United%20States%2C%20food,percent%20of%20the%20food%20supply.

US EPA. "Per- and Polyfluoroalkyl Substances (PFAS) | US EPA," June 6, 2023. www.epa.gov/sdwa/and-polyfluoroalkyl-substances-pfas#:~:text=On%20March%2014%2C%202023%20%2C%20 EPA,known%20as%20GenX%20Chemicals)%2C%20perfluorohexane.

National Cancer Institute. "PFAS Exposure and Risk of Cancer," n.d. dceg.cancer.gov/research/what-we-study/pfas.

"Pharmaceuticals Common in Small Streams in the U.S. | U.S. Geological Survey," March 3, 2019. www.usgs.gov/news/pharmaceuticals-common-small-streams-us.

UNEP - UN Environment Programme. "Plastic Pollution," n.d. www.unep.org/plastic-pollution#:~:text=Plastic%20pollution%20is%20a%20 global,up%20in%20landfills%20or%20dumped.

ReportLinker. "Global Fabric Softeners and Conditioners Market to Reach $19.7 Billion by 2026." *GlobeNewswire News Room*, June 24, 2021. www.globenewswire.com/news-release/2021/06/24/2252358/0/en/Global-Fabric-Softeners-and-Conditioners-Market-to-Reach-19-7-Billion-by-2026.html.

US EPA. "Residential Toilets | US EPA," August 7, 2023. www.epa.gov/watersense/residential-toilets.

Riaz, Tanzeel. "Softeners: Their Application for Leather and Textile Industry." *Uaf-Pk*, December 9, 2018. www.academia.edu/37942840/Softeners_Their_Application_for_Leather_and_Textile_Industry.

Ritchie, Hannah. "Forests and Deforestation." Our World in Data, February 9, 2021. ourworldindata.org/palm-oil.

Royer, Amor. "Copper Toxicity." StatPearls - NCBI Bookshelf, March 27, 2023. www.ncbi.nlm.nih.gov/books/NBK557456.

RTS - Recycle Track Systems. "Food Waste in America in 2023: Statistics & Facts | RTS." Recycle Track Systems, August 14, 2023. www.rts.com/resources/guides/food-waste-america.

———. "Food Waste in America in 2023: Statistics & Facts | RTS." Recycle Track Systems, August 14, 2023. www.rts.com/resources/guides/food-waste-america/#:~:text=The%20average%20American%20family%20of,%241%2C600%20a%20year%20in%20produce.&text=Multiply%20that%20by%20the%20typical,America's%20private%20colleges%20or%20universities.

Federal Register. "Safety and Effectiveness of Consumer Antiseptics; Topical Antimicrobial Drug Products for Over-the-Counter Human Use," September 6, 2016. www.federalregister.gov/documents/2016/09/06/2016-21337/safety-and-effectiveness-of-consumer-antiseptics-topical-antimicrobial-drug-products-for.

ScienceDaily. "Scented Laundry Products Emit Hazardous Chemicals through Dryer Vents," August 11, 2011. www.sciencedaily.com/releases/2011/08/110824091537.htm.

Shaik, Naseemoon, Raghavendra Shanbhog, B. Nandlal, and H. M. Tippeswamy. "Fluoride and Thyroid Function in Children Resident of Naturally Fluoridated Areas Consuming Different Levels of Fluoride in Drinking Water: An Observational Study." *Contemporary Clinical Dentistry* 10, no. 1 (January 1, 2019): 24. doi.org/10.4103/ccd.ccd_108_18.

Collective Fashion Justice. "Shear Destruction — Collective Fashion Justice," n.d. www.collectivefashionjustice.org/shear-destruction.

Srauturier. "Fashion's Water Impacts: The Largest Brands Are Doing the Least - Good On You." Good on You, March 16, 2023. goodonyou.eco/fashions-water-impacts.

Issuu. "The Courage To Change," October 29, 2020. issuu.com/britishbeautycouncil/docs/bbc_20-_20the_20courage_20to_20change_screen_final.

The Economist. "Only 9% of the World's Plastic Is Recycled." *The Economist*, March 12, 2018. www.economist.com/graphic-detail/2018/03/06/only-9-of-the-worlds-plastic-is-recycled?utm_medium=cpc.adword.pd&utm_

source=google&ppccampaignID=18798097116&ppcadID=&utm_campaign=a.22brand_pmax&utm_content=conversion.direct-response.anonymous&gclid=Cj0KCQjwr82iBhCuARIsAO0EAZxvwbw0PudopasU_0PYv558o3DWmQ2a_7g5s-3VVq7hsbUZqQQRk28aAg7XEALw_wcB&gclsrc=aw.ds.

The Groundwater Project. "The Importance of Groundwater - The Groundwater Project," October 27, 2020. gw-project.org/the-importance-of-groundwater.

"The Issue with Tissue," September 13, 2022. www.nrdc.org/resources/issue-tissue.

"The World Counts," n.d. www.theworldcounts.com/challenges/other-products/environmental-impact-of-furniture.

Statista. "Topic: Plastic Waste in the United States," August 16, 2023. www.statista.com/topics/5127/plastic-waste-in-the-united-states/#topicOverview.

Statista. "Topic: Plastic Waste in the United States," August 16, 2023. www.statista.com/topics/5127/plastic-waste-in-the-united-states/#editorsPicks.

Varty, N. "IUCN Red List of Threatened Species: Aniba Rosodora." IUCN Red List of Threatened Species, January 3, 1998. www.iucnredlist.org/species/33958/68966060.

"Vietnam Furniture Market Insights," n.d. www.mordorintelligence.com/industry-reports/vietnam-furniture-market.

US EPA. "Volatile Organic Compounds' Impact on Indoor Air Quality | US EPA," August 15, 2023. www.epa.gov/indoor-air-quality-iaq/volatile-organic-compounds-impact-indoor-air-quality.

Wang, Yufei, and Haifeng Qian. "Phthalates and Their Impacts on Human Health." *Healthcare* 9, no. 5 (May 18, 2021): 603. doi.org/10.3390/healthcare9050603.

Scientific American. "Wipe or Wash? Do Bidets Save Forest and Water Resources?," December 16, 2009. www.scientificamerican.com/article/earth-talks-bidets.

C I R C U M F a U N A. "Wool v Cotton Land Use — C I R C U M F A U N A," n.d. circumfauna.org/wool-v-cotton-land-use.

Xu, Jing, Wangyang Qian, Juying Li, Xiaofei Zhang, Jian He, and Deyang Kong. "Polybrominated Diphenyl Ethers (PBDEs) in Soil and Dust from Plastic Production and Surrounding Areas in Eastern of China." *Environmental Geochemistry and Health* 41, no. 5 (January 28, 2019): 2315–27. doi.org/10.1007/s10653-019-00247-0.

Zhang, Chang, Fang Cui, Guangming Zeng, Min Jiang, Zhongzhu Yang, Zhigang Yu, Meng-Ying Zhu, and Liu-Qing Shen. "Quaternary Ammonium Compounds (QACs): A Review on Occurrence, Fate and Toxicity in the Environment." *Science of the Total Environment* 518–519 (June 1, 2015): 352–62. doi.org/10.1016/j.scitotenv.2015.03.007.

Mango Publishing, established in 2014, publishes an eclectic list of books by diverse authors—both new and established voices—on topics ranging from business, personal growth, women's empowerment, LGBTQ studies, health, and spirituality to history, popular culture, time management, decluttering, lifestyle, mental wellness, aging, and sustainable living. We were named 2019 *and* 2020's #1 fastest growing independent publisher by *Publishers Weekly*. Our success is driven by our main goal, which is to publish high-quality books that will entertain readers as well as make a positive difference in their lives.

Our readers are our most important resource; we value your input, suggestions, and ideas. We'd love to hear from you—after all, we are publishing books for you!

Please stay in touch with us and follow us at:

Facebook: Mango Publishing

Twitter: @MangoPublishing

Instagram: @MangoPublishing

LinkedIn: Mango Publishing

Pinterest: Mango Publishing

Newsletter: mangopublishinggroup.com/newsletter

Join us on Mango's journey to reinvent publishing, one book at a time.